"To all who care about natu[illegible] changing seasons, reading it will be a satisfying experience."

Harper's Magazine

"How much finer our lives would be if we were as attentive to our surroundings as Rachel Peden was to her family, farm, neighbors, and neighborhood. Her ear was equally attuned to rural speech and the grammar of cows, to the creaking of an old house and the chirring of crickets. She rejoiced in every phase of the turning year, every hatching and harvest, every birdsong and blossom, every kind of weather and useful work. Now we can rejoice in the reprinting of this marvelous book."

Scott Russell Sanders, author of
A Conservationist Manifesto (IUP, 2009)

"Before I knew it, the year had passed, and I felt energized, optimistic, and entertained—as if I had spent the period actually in the company of Rachel Peden."

Janice Stillman, editor of
The Old Farmer's Almanac

❋

"Nature springs from these pages. Rachel gently reminds us that we owe the future our past. May this book light a warm fire in the hearts of many eager to somehow counter the speed and urbanization of modern society and culture. Let wild things be wild and do not be afraid."

Tim Grimm, Indiana singer-songwriter

Rural Free

A Farmwife's *Almanac* of Country Living

BY

RACHEL PEDEN

Drawings by Sidonie Coryn

an imprint of
INDIANA UNIVERSITY PRESS
Bloomington & Indianapolis

This book is a publication of

Quarry Books
an imprint of

Indiana University Press
601 North Morton Street
Bloomington, IN 47404-3797 USA

www.iupress.indiana.edu

Telephone orders 800-842-6796
Fax orders 812-855-7931
Orders by e-mail iuporder@indiana.edu

✣ ✣ ✣ ✣ ✣ ✣ ✣ ✣ ✣ ✣ ✣ ✣ ✣ ✣ ✣ ✣ ✣

♾ The paper used in this publication meets the minimum requirements of the American National Standard for Information Sciences—Permanence of Paper for Printed Library Materials, ANSI Z39.48-1992.

Manufactured in the United States of America

Cataloging information is available from the
Library of Congress.
ISBN 978-0-253-22161-2 (pbk.)

2 3 4 5 14 13 12 11

Cover illustration courtesy Library of Congress, Prints and Photographs Division, WPA Poster Collection, LC-USZC2-3742 DLC

CONTENTS

FOREWORD

Time accumulates around our southern Indiana farm in layers. Munching on last year's dried pears, we enjoy this season's crop in bloom. This year's pups from last year's rescued basset hound wrestle in the yard. Our house still rests on the old beams and unmortared foundations originally laid 110 years ago when one of Reverend Livingston's sons walked a little ways down the creek from the log house his father built to a likely spot overlooking the sandy bottom where he would spend his days. It is a brilliant site, on a height of ground with good views and morning sun, tucked on the leeside of low hills that ward off marauding winds. Back in the narrow hollers, mushroom hunters still come upon stone chimneys of a larger community, all tumbled into rubble now, serving only as cairns to mark the passing of human endeavor.

When I dig in my flowerbeds around the house, it's not unusual to turn up signs of earlier residents in sharp bits of coal, the flakes of rusted metal, and the ubiquitous shards of heavy crockery, brown or white or the occasional treasure of robin's egg blue. Some of the shards have the smooth finish of later vessels cast in forms. Others carry traces of the potter's fingers in grooves etched in the local clay as it was thrown on wheels that turned on the power of Richland Creek, just two miles and maybe a hundred years away.

The field below the house shows legacies of our earlier crops in a few patches of oregano and lamb's ear gone wild. There, especially when the soil has just been turned and after a washing rain, even more certain reminders of human persistence in the wide world surface: stone points, pestles, and axes, each fashioned with care and effort and each lost, broken, or discarded in the course of important and vanished purposes thousands of years ago. Under all of these remnants of our best efforts run the elemental currents of water and stone, drop by granule over the eons. This land is riddled with limestone caves, underground rivers, sinks, and seeps that open their secrets only every so often to air and sky. They remind us, if we listen right, that each of us is unique and "not one of us irreplaceable." So says Rachel Peden, who lived not far from here. She knew this country.

For many years Peden wrote about the joys and challenges of country life. Her newspaper columns in the *Indianapolis Star* and the *Muncie Evening Press* intrigued Indiana residents for decades, and her books *Speak to the Earth* (1974), *The Land, the People* (1966), and *Rural Free* (1961) captivated a wider audience around the country. In *Harper's Magazine,* Katherine Gauss Jackson called reading *Rural Free* "an evocative and satisfying experience," a book "very much in the midst of life." The world Peden describes was fading then and may feel old-fashioned to us now, as when she describes walking her petition down gravel roads for a neighbor's signature, having unexpected visitors drop by her kitchen for a cup of coffee, or watching when two men stop their trucks in the middle of the road for a leisurely catch-up on local news, never needing to hurry for oncoming traffic. The rhythms and patterns of country life that she measures convey a compelling sense of security, home, place, purpose. In short quips and meditations, she gives us a backdrop of "simple available riches," against which she sets the drama of glorious sunsets, the antics of unpredictable farm animals (human and otherwise), poetic gems from hill farmers, and her own "vacations in the blackberry patch." Peden came to her Monroe County land on the back of a horse-drawn wagon, forced out of a beloved homeplace by the double burden of a heavy mortgage and the expensive prospect of mechanizing the farm. Coming as she does at this cusp of the old farming ways and the new gives her a special voice to speak to our own time—when people look to bring back the best of the old ways into our fast-paced, often rootless lives. This twentieth-century "fireplace philosopher" offers much to inform our own visions of the possible. Her record of a time gone by can help us to reclaim the best features of the community she describes, more diverse perhaps in membership and opportunity than hers but with a similar sense of mutual reliance, appreciation, and forbearance.

When Peden wrote *Rural Free,* the United States was in full-scale urban sprawl, with the number of farms shrinking. Today that trend seems to be leveling off. Although good farmland continues to be lost to shopping areas and housing developments, nationwide the number of farms is growing. The 2007 United States census counted a 13 percent increase in farms from 2002, to more than 2.2 million farms total. Most of these new farms tend to be smaller and yield more diverse products. Many of them are owned and operated by "retired" people, and the majority require an off-farm income to survive. The number of farms run by women increased 30 percent in that five-year period, and Hispanic, Native American, and African American farmers are also increasing in number. Not only are people moving to farms, but urbanites want to connect with country life. Farmers' markets have grown more than ten-fold from about 400 in 1970 to 4,500 in 2007. Community-supported agriculture ventures that allow townspeople to secure a share of the season's harvest have also been growing by leaps and bounds. The Agricultural Census shows that currently more than 12,500 farms market directly to customers through CSAs. Many of those farms offer days when customers can visit to pitch in their labor, picnic, play with the chickens, hike, or meet other customers at potlucks and barbeques. And more and more townspeople proudly give over part of their yard to a vegetable patch.

Growing your own and relying on local seem less elitist and more a matter of commonsense security than they might have before. Salaries and housing markets are stagnant at best, refocusing people all over the nation from their move to the next McMansion to seeing instead where they have already arrived. What makes this place where we find ourselves unique, worthwhile, a place to stay? What makes this *here,* and what from this place can make me more *me*? As Peden says of her own time, fifty years ago:

> This generation is beset in its daily life by fears, angers, blockings of its most passionately cherished hopes, by loneliness, and by the need of solitude.
>
> The opportunity of looking intently into some small portion of his natural environment, or the experience of being responsible for the survival of some part of it, it enables a man to see his own place in the world with greater clarity, and perhaps with greater compassion.
>
> It is only when he sees how everything is vital to the pattern and fits into it that he achieves a kind of refuge for his own buffeted spirit. It is not a hermit hiding place—the modern world has no place for a hermit—it is a kind of spiritual symbiosis. It's the beginning of the science of kindness.

Even before most of us realized home would become a rare commodity, Peden was urging us to embrace the place in which we find ourselves. She recognized that respect for the environment is inseparable from respect for our community and ourselves. She saw that we are all, as John Donne said, "a part of the main," essential actors in our collective sustainability. She was pointing us home.

When Peden talks of spiritual symbiosis, it is in the context of an entire world inspirited. Common crickets and calves convey ancient continuities. The rituals of school buses, solitary milking, family seed sorting, and the collective harvesting of crops ground our complex world in timeless patterns and relations. She describes a children's swimming hole repurposed for an immersion baptism the day before a young man goes off to military service. She grants us a funny rescue for her husband nearly buried alive in a silo accident. She gives us her cranky, rickety house transformed into a cheery home. Her Currier and Ives mornings almost outglow the other particulars she threads throughout—that crops fail, sometimes horses are thin, mortgages can go unpaid, and barns burn down. She reminds us gently that we cannot withdraw from the challenges of life. We must know "that the everyday world is always there, to be confronted, and that there is strength with which to confront it." Her stories are much like the child's hand she describes, "opened and held palm up to show that nothing is concealed in it," "the naked facts" of habits and holdings revealed. The inevitable questions of existence, the meaning of life, our own individual purposes are unavoidable, she indicates. But if we listen closely and long enough, she says, we will hear "fumbling, incomplete, and elusive" answers that come from the whole farm as its collective wisdom, a "reassurance carried from an ancient civilization into our modern one" that no one is alone, no one outside the web of life. And when it seems we have run afoul of the natural way, it's only that we have not understood how this piece, too, is part of the whole. We matter and we do not. That nothing depends on us for the world to go on is both a freeing and a humbling philosophy. Or, as her husband says, "lots of things solve themselves if you just let 'em alone."

We should all hope to be invited into a kitchen like Rachel's to share such thoughts. There we might get fried rabbit with creamy gravy on mounds of rice with buttered new turnips, crisp pickles, and chocolate fudge cake for dessert. Even a simple cup of coffee and a doughnut at the kitchen table over talk of the similarities between people and livestock or the disparities between humanity's grand plans and nature's inexorable intentions would be time well spent and pleasurable. Such is Peden's witness to the "gospel of beauty in the commonplace."

JENNIFER META ROBINSON
Greene County, Indiana

ACKNOWLEDGMENT

This book began life through columns written for the *Indianapolis Star* and the *Muncie Evening Press*. The author is grateful to the editors of these newspapers for generous permission to use the material. The author is grateful to E. C. P. and to A. C., who said, "Certainly you can write a book"; and to G. M. L., who proofread the manuscript; and to column readers. Also to the neighbors for permission to turn the farmlight briefly and honestly, lovingly, on them. And the author is grateful to you, the reader; for without you this book would be like the hypothetical falling tree that made no sound because it fell where there were no ears to hear.

SEPTEMBER

[SEPTEMBER]

"YOU can tell the days are getting shorter, can't you?" said Dick at breakfast this morning.

From where he sits at the kitchen table he can look through the west window and watch the year go across the farm in a familiar pattern, never identically repeated. It is made up of many small things, and there is always something in it that seems like news. It may be the black-cowled cardinal lunching on winter-gray seeds in a Queen Anne's lace plant in the garden; it may be the awesome orange-pink glow spread on the upsloping west pasture from sunrise at the other side of the house; it may be the sudden flock of red-winged blackbirds, pausing in the harp-shaped ash tree on the hillside, in which are still the remnants of the tree house the children built there ten years ago. Sometimes it is a pack of wild dogs furtively crossing the rim of the hill toward the woods; sometimes a red fox walking there alone, proud and unafraid, coveting the tender flesh of the banties that are safe in the bantyhouse near the barn.

Putting toast into the oven so the butter would melt into it, and listening for upstairs sounds made by Joe, a senior in high school this year, and Carol, in junior high, I stopped to look out at the September-shortened morning. Above the hilltop I could see the tip of the young maple by the sink-hole. Its leaves seemed not actually yellow, but a little less green than they had been yesterday.

"Yes," I agreed, "summer's over."

The morning had a sense of beginning and end overlapping a little bulkily like the seam on a milk bucket.

A farm year has, actually, no beginning and no end. Like an old-fashioned roller towel, it begins any place a farmer can get hold of it. Moving to a new farm, he tries to get there by the first of March, to start his spring work. Or he goes in the fall, having sold all the corn and hay, machinery and livestock he can spare, and leaving the last of his garden to its weedy fate. But for farmers who have children in school, the year begins when school begins. Then the big orange-colored school bus scurries like a beetle along all the country roads, pauses long enough to gulp down the children, and is gone. Behind it a new routine settles down over the farm.

The beginning of a year

I took the toast out of the oven, laid it beside a little rick of bacon and two eggs on Dick's plate, and poured myself half a cup of coffee. Usually we eat a leisurely breakfast before the children come down. This morning they came just as I took the first sip of coffee. It will be that way from now on. School has begun; it's the beginning of a new year.

Presently the year will begin to nibble away its accumulations, but now, for the flicker of a sunlit moment, it pauses to estimate its harvest.

When the children went down to the bus, sunlight was bright and diffused as if pressed out from under a heavy lid, but in the west the moon hung, worn and transparent as the last melting sliver of candy sucker in a child's mouth. Across the road the hillside lay fog-silvered and unawakened. Foxtail grass, which had finally overtaken the watermelons at the garden, wore a headdress of silvery dew in its abundant rough hair.

By noon the dry clothes danced and fluttered on the clothesline and wind turned the silver side of maple leaves out, which means rain. Grasshoppers and black and gold tiger swallowtails moved casually out of my way as I went along the line, taking down the sweetly fragrant clothes. At my feet sourgrass offered its last crop of tiny gold cups.

In weather "hot as the hinges" all week, the summery beauty of growing corn has been translated into a ripening harvest. Weeks ago the brown came up on the stalks, but the long, warped, sharp-edged blades are still green and there is juice in the yellow ears, encased in their rough, corded husks. There is juice, too, in the tall stalks, and when

Dick brings in a wagonload of cut corn at evening, every animal on the place welcomes him. Babe and Lady and their half-grown colts nicker from the fence, red pigs suddenly appear, squealing delight. Cattle bawl as they run toward the wagon.

In the chicken lot now Rosemary, the speckled Bantam hen, still chaperones her little brood with the tenderness she gave them in their fluffy chickhood. Now that they need her less, however, they roam farther from her and Rosemary has to look carefully to know whether it is one of her own, or a hungry sparrow, that has flown down uninvited to share her breakfast.

THIS WEEK the diminishing moon is a golden watermelon cut into thirds. On the golden flesh you can see the shadowy golden seed pockets, and the golden seeds, too; and the moon's burning light poured down on the ground is a golden juice.

Nights are marked by a steady humming spread on the air like a thick blanket. Only by listening acutely can you detect separate voices in the humming. There is the challenge and rebuttal of katydids in ceaseless debate; the short, quick shrillness of a cricket; the shaken vibrato of cicada, jarfly, and other insects that hum in the night. All these blend into a chorus from which occasionally you may hear the quaver of the gray screech owl, the rare cello of a big bullfrog at the pond back of the barn, or the dramatic interception of some dark bird's sudden outcry. The whole orchestration is not loud enough to waken a sleeper, but if you happen to be already awake and listening, you are aware that a great amount of sound is harmoniously blended to make "the quiet of the country."

THIS IS the canning moon, and you don't have to be an Indian to recognize its signs.

Farm kitchens are fertile with bushel baskets of green beans, tomatoes, cucumbers, cabbage, corn, or flat baskets of purple or white grapes, all to be got ready for freezer or pressure canner.

A farmer bringing in the morning milk is likely to find the kitchen floor carpeted with corn husks, and from the hog lot comes the happy, gluttonous chomping of hogs stowing away the milky, close-shaved cobs. By these signs the farmer knows freezer and cellar are waxing full.

At noon he will know it even more. He can tell by whatever is in the big, fluted, white ironstone dish just what phase the canning moon is in. In the sign of the green bean, dinner will be green beans and bread and butter. The next day, crescent in the corn phase, dinner will be roasting ears and bread and butter. The third day, in the sign of succotash, dinner includes sliced tomatoes. After so long without pie, a farmer waxes hungry. Even so, he fares better in the canning moon than in the house-cleaning moon, in which there is no assurance even of bread and butter to eat.

"AND BEHOLD," says the Book of Exodus, "the glory of the Lord appeared in a cloud." It was probably a sunset, of the kind that impels a farm child to run down from the barn and ask the mother breathlessly: "Oh, honey, have you noticed the sunset tonight?"

This family sharing of simple available riches is one of the rewards of living on a farm.

Sunset often is spectacular. Perhaps at the end of a day the human need of drama is greatest. Perhaps the glory of the sunset is to suggest that there is more order than chaos in the natural world.

The colors vary from lakes of fire to long streaks of red fading to orange, to salmon pink, finally to palest yellow, becoming, at the last, merely the color of sky. Sometimes the pictures are classic, sometimes modernistic streaks of color and guesswork. The texture appears to be foam, smoke, water, fire, rock, great banks of snow, wind.

In swift violent scenes horses plunge and rear; chariots race to battle; great armies march across mountains edged with fire; whole nations migrate recklessly; great ships are tossed in mighty oceans. In the glory of the sunset one sees familiar sights: forests; fields of ripe wheat blown in the wind; corn in shocks; pointed towers; sprawling cities. Violence erases the quiet things: volcanoes rising and smoking and giving way to fierce, unassailable cliffs overlooking quiet, pale-green seas. One picture merges into another so that a watcher can hardly point it out to a companion before it has changed. Even as one mentions the ship with flying flags and waves curling against its sides, it has become a deserted farm, with abandoned sheds leaning tiredly toward a weed-filled garden.

Sometimes the sun dominates the whole majestic scene, a brilliant globe of transluscent, burning red rolling down behind a brown hill. Sometimes the sun is merely the opalescent pink swish of a giant brush across the sky.

When the last stain has faded, the watcher is abruptly aware of the contrasting darkness on the ground. Everyone has seen in the sunset something peculiarly expressive of his own experience and beliefs. The glory of sunset is reflected against his spirit, even as its light on his uplifted face.

ARROWHEADS were in bloom in the shallow pond at the top of the hill when I passed there today, and I remembered vividly the first time we found them. Joe was eleven then; the farm was a source of continuous delight. He came down from the cornfield, excitedly offering me two stalks fourteen inches long and one tightly closed triangular bud a fourth as big as a spool of thread. Each stalk had one leaf, heavily ribbed and sharply pointed at its three corners. Except for the long slit from stem to leaf center, the deeply cupped leaves would have held water.

"Is it a pond lily?" asked the eager boy.

I didn't know either, so we got out the plant book and identified it.

It was an arrowhead, a member of the water-plantain family. It thrives, the book said, in open ponds and marshes of sluggish streams. It blooms with a pure-white three-petaled flower on a stalk a foot or more taller than the leaf stalk. If nothing interferes with its development, the flower gives place to a small one-seeded capsulelike fruit.

When we moved to this farm, the pond was no more than a swag in the field where water stood briefly after a heavy rain. It dried out too quickly to sustain any kind of aquatic plant or fish. Cattle wading in to drink eventually sealed the leak, and the pond became permanent. But where did the arrowheads come from?

Nature's means of getting her seeds around are wind and water, the feet and craws of birds, the feet and droppings of grazing animals. A bird's foot could have carried seeds of arrowhead. But why would a water bird have stopped there, where water stayed so briefly?

Arrowheads spread, almost too willingly, by underground stems and under favorable conditions develop large, starchy, potatolike tubers, which ducks eat with considerable relish. Long ago, the plant book said, the tubers were roasted or boiled and eaten by Algonquin Indians.

At this point our research took on an exciting personal interest. On this farm, and on Alec Ulmet's farm across the hollow, Indians walked long ago, had campfires, signaled from one hilltop to another, and lost flint arrowheads in the ground. Every year, when the fields are plowed, we find a few more of these sharp-pointed weapons of long ago. Perhaps at that time there was a pond in the same place, and there the Indians roasted their tubers, overlooking a few small ones deep in the waterlogged ground. Drought held the tubers dormant. Later, when enough water to support plant life collected and stayed, the tubers sprouted. Could this be the explanation? If it is, how do such plants reconcile themselves to the long, patient time of waiting? How do they know when to start again? And where do young boys get that shining look on their faces and that voice in which eagerness has the sound of lifting wings?

THE COMMON CRICKET is small, dark-brown, and numerous—also noisy. "Ye-up, ye-up, ye-up," he sings harshly. His voice sounds like a burlap sack filled with corn being dragged bumpily over a stony road. His ambition is to get into the house and hide, there to sing until winter overtakes him or he gets stepped on.

The song of the cricket is said to instill a note of content into a house. It does, of course, if you are the cricket. It does also, to some extent, if you consider that the voice of the cricket in today's complex world is just the same as it was in an earlier American era, or even in a much more remote civilization. The same voice undoubtedly was heard by Indian women picking off the small maize ears they had produced by poking a hole in the ground with a pointed stick and dropping in a dead fish along with the precious seed. The same voice sang much later, too, when farmers went out on frosty autumn days with a hickory shucking peg and a team and wagon to gather a load of corn before the ground thawed. The team was trained to move forward a short distance when the farmer said "Get up" and stop when he said

"Whoa!" The cricket's voice is just the same now that farmers use a savage two-row picker traveling swiftly in third gear to pick corn; and if the farmer gets careless, the picker will snatch off his arm as impersonally as it would an ear of corn.

The voice of the cricket is a theme of reassurance carried from an ancient civilization into our modern one. If I ever have to dive for survival into an underground shelter, I hope there will be a cricket familiarly "ye-up, ye-up, ye-upping" there.

THE WIND seems like autumn wind now, as it scatters the first torn-off yellowed leaves.

Even before the leaves began putting on their red and gold, cattle began putting on their winter wear. Washed to glossiness by late rains, their hair has thickened and lengthened. Long before walnuts drop on the ground and dark wild grapes have been eaten by birds, farm animals are dressed for winter.

I looked out into the chilly gray morning ribbed with rain and saw the cattle placidly grazing in the pasture.

"Cows don't mind the rain at all, do they?" I said.

"No, they don't," agreed Dick. "It doesn't even get down to their skin. You can part the hair, and the skin is perfectly dry. When the water strikes the hairs, they come together and make little pointed clusters. The clusters overlap like shingles on a roof and carry the water away."

What "poor, cold things"?

The longer the cows stood in the rain, the darker and shinier their sides looked.

"In winter," continued Dick, "the water freezes and clings to their sides in little icicles, snow lays on their backs, and people say: 'The poor, cold things!' But the truth is, if any heat were getting out through their skins, the snow would melt from their backs and the icicles would drop off, too. Cattle have perfect insulation."

He poured himself a glass of milk and drank it, then poured cream into the glass and filled it with coffee.

"And I've never been able to find any moist skin on a newborn calf, either," he added. "It's simply marvelous. I have parted a calf's hair and rubbed the skin with a dry finger and never could get off any moisture."

The thought drops a light covering of awe over a commonplace thing. Nature has a great number of these small, remarkable surprises. Discovering them makes farming interesting and mystifying, for they give one always the feeling that something momentous is about to be revealed.

MID-SEPTEMBER now, and the hummingbird, as if suddenly aware how late it is, dips deeply for the scanty nectar from late, flame-colored sultanas and pink althea blooms.

Autumn insects sing a sleepy verse now. It sounds old, worn thin and comfortable, like the knees of old overalls. The cricket seeks a sheltered crack from which to fiddle his sleepy squeakings. His songs are more bearable than his first, too-vigorous verses.

Along the side roads the bright gold of thin-leafed wild sunflowers gleams from its dust covering and attracts the eye as quickly as mention of easy money. Purple ironweed is diminishing in the pastures; thistles are down to their last silken tassels; goldenrod pours its heap of raw gold into the general fund.

From tall grass and weeds at the roadside, wild asters are ready to stop passers-by with their simple loveliness. All summer the stalks grew straight and tall, bearing unpretentious foliage. Then, when the flower buds came on, the tall stalks knelt toward the road. Now, out of a fringelike lacy mass of leaf they offer their particular purple. It is neither blatant nor pompous, but sudden, compassionate, and toned down as if generous amounts of autumn's late, luminous moonlight had been poured over it and absorbed.

WHERE a barn burned one wet morning a few years ago, some of the foundation stones remain, making mowing difficult. Yesterday, trying to avoid a big stone, Dick inadvertently hooked the cutter bar against a smaller stone. It dislodged the stone, revealing a bumblebees' nest underneath, and brought the bees out in an angry rush. Dick got away unstung, but Rose, the farm collie, was less fortunate. A bee became entangled in her long black and white hair and stung her before Dick could get it out.

Later he said: "I think I'll take a little bunch of straw and go down there and. . . ."

"Oh, I wouldn't kill them," I protested.

"Well, I don't think we need a hive of bumblebees here," he insisted.

One reason I spoke up for the bees is that every summer, when the calico-pink spiraea blooms by the kitchen walk, a bumblebee visits it. The bush spreads out, pink-sprigged like an old-fashioned full skirt, and the bee's big incongruous body, gorgeously black and gold velvet, makes an unforgettable moment of beauty as he pauses there.

I admire bumblebees also because they appear to defy science. According to science, they are so constructed that it is impossible for them to get themselves air-borne. Being uninformed of this fact, bumblebees go right ahead and fly as they please.

In places where there is no winter they live all year. In chillier climates, such as this one, the workers and drones die in the fall, having first made sure of bumblebee posterity by fertilizing a number of young queens and providing them safe housing for the winter. The queens hibernate through the winter and in spring hunt up some cozy spot like a chipmunk's abandoned foyer or a stone warmed by sun and concealed by weeds, and there deposit eggs.

If beauty and science-contradiction are not enough to get this member of the Hymenoptera clan off the hook, there is another good plea, especially for farm bumblebees. They pollinate red clover, which nourishes cattle and helps hold a farmer's restless, perishable soil on the rocky bones of a hill farm.

In this indirect way bumblebees make their contribution to a civilization which, having devised marvelous machinery for flying, sometimes seems bent on using it to destroy its whole family and all its wild neighbors besides.

ALMOST every farm has its personal sinkhole in the back pasture or woods where a farmer dumps the debris of his changing life. It is the final destination of the worn-out washing machine or the broken-down gas stove, the obsolete cream separator and brooder stove, the sacks of glass cans and empty bottles, the leaky milk bucket, and other household utensils beyond use.

Farmers are reluctant to abandon anything irretrievably. They say it might "come in handy sometime" and let everything accumulate in tool shed, woodhouse, or barn until finally a Time comes.

"Well," Dick always says when the Time comes, "if we need it, we know where it is and we can always go get it out." The sinkhole is ringed with items abandoned on this comforting policy. I wonder how the ancient farm women made out when the Time came to the middens.

The policy of banishment with perpetual right of recall was justified last week when my neighbor Sairy Brown stopped by with Esther, her guest from Texas.

We walked up to the woods.

Esther is that rare Texan who doesn't believe Texas has

everything. The fact that she was born in Indiana has some bearing on this attitude probably. As we walked through the luxurious leaf-strewn woods, Esther kept stopping to kneel down and run her fingers hungrily through the deep leaf mould. She yearned to take some home to Texas, but we had nothing to carry it in. Esther's close-fitting blue skirt had no lap; her white rayon blouse had only one pocket. Her hands were already filled with ripe papaws.

Eventually we came to the sinkhole and there beheld a great multitude of treasure: old bottomless buckets, cracked crocks, a large dishpan, rusted tin cans. Richest treasure of all was a white enameled vessel, three-gallon capacity. Its bottom was rusted and leaky but could be covered with mulberry leaves. Its handle was gone, but it could be clasped around its urn-shaped middle and carried in one's arms. Blissfully, Esther filled it with leaf mould. On the way down to the house she set it down only when she had to crawl under a barbed wire fence.

When she has used up all the leaf mould around her plants and vines back in Texas, she said, she will fill the old vessel with philodendron and give it a place of honor in the bathroom window. Life at the farm sinkhole is not without hope.

THE NEW implement shed now approaches the triumphant hour when machinery can be stored in it. Considering the present cost of lumber, this is no small accomplishment. The joy is diminished for Dick because he realizes that the new oak and poplar boards, bought for this shed but laid aside by him "because they are just too nice to saw up" will soon have to go into the shed's walls or be replaced by others, equally "nice" and just as expensive.

"This shed will cost four times as much as it ought to," said Ralph, the neighbor who is building it, "because of the lumber he's saved back. If I want to use a good board, I have to take it when he isn't looking."

"This is the last building I'm ever going to build," lamented Dick. "Using those good boards is even harder than killing pets."

Nobody mentioned it just then, but someone could have reminded him of the two expensive gates he bought and stored in the lower barn because they were "too good to use." They were still there, just as good, when the barn burned.

A Psalm of Grapes

BEHOLD, this is the season of grapes.

Sing of Concords hanging frosty-purple, sweet, and sun-warmed in heavy clusters from the vines, whereof the leaves are beginning to turn brown at the edges, presently to curl up, wither, and drop off.

The leaves that remain green on top and pale-gray like suède underneath are like unto those already picked and placed on top of big stone jars filled with cucumber pickles, with an old china plate and a clean stone on top of them.

Sing of white Niagaras, round and sweet and possessing an ineffable bouquet which is always the same, year after year, as if the same identical grapes returned every year to the same vine, and likewise the same hungry wasps. Sing of the young farm boy climbing the slim black walnut tree that the vine had climbed earlier, it to produce the grapes and he to possess them.

Sing of grapes hovered over by wasps, hornets, yellow jackets, and honeybees which, having already gathered the flavors of clover, alfalfa, appleblossom, elderberry, peach, and the juice of bruised Seckel pear, now ask for only one more, grape. Sing of grape-eaters, how some suck the juice out of the hull and throw away the hull and also the seeds; and some laboriously separate and spit out the seeds; but the wisest among them bite down and swallow first the juice, then hull, seed, and pulp.

Sing of women making grape jelly, watching for the "rolling, tumbling boil that cannot be stirred down" and the juice sheeting from the spoon when the jelly is ready to be taken off the fire; of the filled glasses set in a window where the light can shine through and show the beauty thereof. Sing of paraffin, melted in a small coffeepot, to cover.

The time of grapes is a time of goodness, richness, and hope. For, behold, did not Jeremiah, the ancient prophet of doom, yet comfort his people with the promise, "Thou shalt yet plant vines upon the mountains. The planters shall plant and shall eat them as common things."

Behold, the authenticity of autumn is made fast in the fragrance of grapes.

FOR DAYS now our dinner-table conversation has been flavored with plans for silo-filling.

This year ours will be the first silo filled in this community because we have the earliest corn. It must be done on Saturday, so that Joe can be here, for this is one of the most exciting events of the farm year. Farmers dread it, in the same way a club woman dreads reading a club paper, but it is a social event no farmer wants to miss.

Preparations have been made; the blower mounted beside the silo with its big pipe hanging over the rim; the cornfield opened by cutting away the first rows of corn, thereby clearing passage for a tractor to come in, pulling a one-row field chopper and rubber-tired wagon with boxlike bed. The amount of money invested in the machinery necessary to fill a modern thirty-foot silo would have bought four farms ten years ago.

Carol and I planned dinner for ten men. (Ten years ago, when the corn had to be cut by hand and hauled in "on the long stalk" on wagons, the work would have required twenty men and dinner would have required at least four women cooks.)

My neighbors offered to help, and sent garden offerings. Elizabeth sent late green beans and roasting ears; Grace sent Wealthy apples; Mary dressed two guineas for roasting. There will also be fried fish, potato salad, coleslaw, sliced tomatoes red and yellow, cottage cheese, sliced onions in vinegar, baked beans, celery, baked macaroni and cheese, homemade rolls with butter and blackberry jam, glazed sweet potatoes, iced tea, hot coffee, three kinds of cookies, lime sherbet with ginger ale poured over it, and pumpkin pie.

Dick planned quietly; Carol and I used pencil and paper. But Joe planned aloud, rapturously and everywhere. And at the family supper table suddenly his planning broke out into a chant, and nobody interrupted. "Oh, I love it! I wish we could have fifty neighbors here. I love the jokes and the talk and the tricks they play on each other, and the noise of the machinery. I'll get to drive the tractor that pulls the chopper. I love the sound of the tractors going to the field and coming back with the wagons full of chopped corn, and the cows bawling when they smell it, and the ponies running up to the fence, and the little stallion nickering when I drive past and whistle to him.

"I love the loudness of the blower at the silo so that if you want to tell somebody something you have to lean over and yell at him, and somebody tramping down silage and his head showing above the top when it's getting close to the last door—but it'll settle a door and a half that night and we can run in a little more. And shutting down the machinery when it's time to come in to dinner, everybody laughing and talking and a little embarrassed and anxious to get to dinner. I love the sweat and dust on your face and the chopped corn down your back and in your pockets. Jerry Bolinger and Dick Francis both asked to come, and Warren will bring Dave, I suppose. Warren's always so full of life. And Fred Dutton, he's a dear old man, I could just hug him. And Carr 'gumming it' at the table, he can ask the blessing. And late in the afternoon we can come in and have ice-cold watermelon out in the yard and everyone can sit around and talk and rest. I love it because when the silo's filled you know you've accomplished something important, and you wish you had to do it all again the next day. Oh, I just can't hardly wait!"

NEW harvesting machinery seems, at first, expensive, or as Carr Stanger says, "a little salty." But when enough farmers take up the new way, the cost goes down; and then more farmers adopt the new way until finally it is difficult to get enough of a crew to harvest a crop in the old way.

That's what happened to Alva Jacobs Friday. He was going to fill his silo the old way, cutting the long stalks with corn knives and running them through a cutter at the silo. It would take twenty men. Dick had been asked to come "singlehanded," that is, without a tractor, because there would be six tractors there from farms nearer Alva's than ours is.

Dick was driving, and I went along so I could bring the car back and have it that day.

At the crossroads we met Warren Fyffe in his truck and stopped for those few minutes of visiting that farmers consider more important than anything else.

"Well, don't work too hard," warned Warren finally as we started our separate ways. "I wish I could go with you."

"Where is he going?" I asked.

"Oh, just up to the church to mow weeds."

"Well then, why can't he go to the silo-filling instead?"

"Oh," explained Dick, driving along at twenty miles an hour, "he don't really want to. He just wishes he did want to."

One thing about driving twenty miles an hour, you don't miss much. "Now, that is the kind of road I like," said Dick, "when you can just pull up broadside of a neighbor and you're not in the way of anything."

It is five miles from our farm to Jacobs's. The way goes past Modesto, past Jane Love's old farm, and over the Moll de Lapp bridge across Beanblossom creek.

"It's not often anymore you find a road that's got a strip of grass growing right down the middle," commented Dick. We passed Emsley Fyffe's house, set high up on a thin-soiled hill as steep as the licked-off rounded top of an ice cream cone. Edith was at her henhouse at the foot of the hill, feeding her glistening white hens, and we waved.

Horseweeds, now gone to seed, stood close enough that I could almost reach out and pick them from the car window. Birds flew out of the trees, shaking down raindrops from last night's rain. Weeds and leaves gleamed wetly under the gathering strength of the morning sun, but there were shady narrow places where limbs from trees on one side of the road leaned out to touch leaves on the other side, and little branches flicked the car's radio aerial into a constant twittering. Dick sounded the horn as we drove into a narrow covered bridge, and the plank floor rumbled as we went across it. We emerged in a sharp curve and passed a cornfield where the crop this year has been a complete failure. The pale, thin stalks had no ears on them. It was a creek-bottom farm that had been flooded until late in the spring and remained too wet to break until midsummer and, as Dick suggested, "probably never saw a load of manure in its life."

The road goes down through woods and hills, follows Beanblossom creek, wades through mudholes, and passes numerous farms, some better than others. It must be difficult for the town-working farmers who make the daily trip to town over that road, a trip that is necessary in order to subsidize the skimpy farm income, so that they can continue to live on the farms.

CHIPMUNKS have had things pretty much their way around here this summer, but this week their peace of mind has been shattered by the arrival of an uninvited cat. Yellow, thin, and gentle, she flatters people by rubbing against their legs, mewing, and looking up with limpid, lovely yellow eyes. She has obviously been a pet, dumped along the road because she was no longer wanted. It's a common occurence along a farm road.

I feed her secretly, hoping she will leave. Dick also feeds her, but not so secretly, since he has to get the food, except fresh milk, from the kitchen.

The chipmunks make no bones about her being unwelcome. They stand in the back yard and protest in an angry, unmusical "chip, chip, chip!"

The yellow cat meantime walks about kindly, rubbing against the legs of cat-allergic guests and purring softly, pretending not to notice the chipmunks. She brings us presents of dead mice and rats, leaving them, like May baskets, beside the front door.

The unreconciled chipmunks have many small holes into which they can dive for safety. There is one by the seedling peach tree, one by the white-rose bush, one in the zinnia bed. A very small hole will suffice if it is in the right place at the right time. But the chipmunks do not like this holey life. They like freedom above ground. Who doesn't, being accustomed to it?

The chipmunks and I have watched each other all summer. We are friends. That's what made this morning's performance really poignant. When I came out to the back porch, the mother chipmunk came and stood on a low brick wall that encloses a flower bed. She stood up ramrod straight on her hind legs. Her front legs were bent at the elbow and held close against the fawn-colored, soft fur of her underbody. She laid one little hand upon the other and just stood there, very still, regarding me respectfully out of her dark, oil-colored, oval eyes.

At the sound of the typewriter she assumed a look of rapt interest. Now and then, in her concentration, she forgot her hands and they dropped down, apart, and she retrieved them with a quick jerk, laying them again one on top of the other. She looked calm, but her tenseness was betrayed by the way her little short ears pointed straight up from the corners of her head like football goal posts, and by the violent quiver that overtook her whole striped brown body when someone in the house let a door slam.

She looked exactly as a petitioner should look when presenting an important petition.

I do hope the yellow cat will not put on a competing performance. Life is complex enough, without having to judge a talent show.

YESTERDAY morning I went up to the woods to see about the papaws.

"Aw pshaw," exclaimed a crow from the hedge as I walked up the hill, "aw pshaw!" He flew across the field to join a noisy crow conference in the woods, where by this time leaves are touched with yellow and pink.

Grass was short on the hard ground; lespedeza was in bright, small, purple bloom. "The boasted Ladino has deserted farmers now," said Dick, "but lespedeza and the ever-faithful orchard grass, a farmer's best friend in time of drought, are still supplying pasture."

Grasshoppers leaped out of my path as if shot out. At a place lately visited by cattle, two hard-shelled black tumblebugs were rolling their ball uphill. From the locust trees, small yellow ovals were blowing down. Walnuts were shedding their narrow, brown, crackling leaves. Near the tulip poplar there was a mechanical-sounding whir as a black and white striped chicken hawk parted angry company with a crow. Small birds are less numerous now, as if nature had sifted the birds through a coarse sifter and only the larger ones remained.

At the first step across the threshold of the woods I felt

the welcome coolness, and along the narrow tractor road the rustle of dry leaves sounded like soft footsteps. At the papaw grove I saw that moles had tunneled toward the bushes, pushing the old leaves, gray as a hornets' nest, into long ridges.

There were three large papaws in the taller tree, but they did not fall when I shook the tree gently, so I knew they were not ripe enough to take. The papaw leaves are beginning to turn yellow, like big thin tobacco leaves, but the fruit needs more time. Papaws should be tree-ripened, yellow, flecked with brown. The inside will then be soft as a custard, yellow as rich cream. A couple of papaws on the kitchen table will cause an eagerly sniffing visitor to exclaim: "Papaws! Where did you get them?"

YOU ARE not alone, going to a woods now, even though nobody walks with you. It is a quieting journey, full of discovery. In sharp contrast you step from the sunlit area of bright goldenrod and lavender thistle at the field's edge into the cool dimness of the woods, where patches of light and shadow dapple the leaf-strewn path.

Beechnuts and their burry, three-pronged pods are dropping, though not yet dark brown. Nearby I discovered some clusters of tiny orange-colored fungi on wirelike, drying stems. The flat tops, varying in size from a shoelace hole to a quarter, are also dry and ribbed like souvenir party parasols.

Side-stepping the small round burrs that were eager to

go home with me, I walked into a spiderweb. It was almost impossible to avoid them. For the most part, spiders now weaving in the woods use the traditional pattern, a web anchored by long filmy cables and containing at the center a delicate threadwork doily. The smaller webs I saw were presided over by small, humped-up spiders, but near the beech tree were half a dozen larger webs, some five feet long. Sunlight poured through their large center doilies like milk through a strainer. I had stopped to examine one web, twelve by eighteen inches, and noticed a small green bug that paused on a delicate cable. Instantly, from no place, it seemed, the spider was upon him. She was a huge spider, black, orange, and chartreuse. There was no contest, but almost immediately there was no green bug, either. The spider remained on the web long enough for me to observe her beauty and the cleverness of her hiding place.

She had pulled three beechleaves together into a kind of funnel in which she concealed herself. She had eight legs; the leg-joint next to her orange-colored bodice was orange, the two lower joints were alternately black and white. Her body was as large in diameter as a shirt button and exquisitely structured. On a pincushion-shaped background, spots of chartreuse were patterned like the figures of people, vases, and flowering vines in ancient Egyptian carvings.

As I watched her, I was aware that the silence of the woods was filled with the conversation of birds. They were not singing; they whirred, cawed, clucked, clicked, mewed, chipped, and muttered. The only one I could identify was the crow. But I was deeply conscious that I was not alone.

In the woods around me were myriads of lives, every one important and significant unto itself, and not one of us irreplaceable.

The fairy-tale frog

On a summer evening, when the kitchen light is burning, there are always multitudes of drab-winged moths fumbling against the window, trying to get in and die in ecstacy against the hot light bulb. You hardly notice them, because they always come.

One rainy evening this week, as I was washing the supper dishes, my attention was suddenly pulled to the window by the realization that something unusual was happening there.

A tree toad was walking energetically over the damp outside of the pane. He was tiny, a miniature, fairy-tale frog. (Some people do call them tree frogs.) His body was no larger than the first joint of an adult's little finger. His long hind legs, if stretched out straight, might have been an inch and a half long, and his front legs were no more than a third of that. Fascinating were his toes—five on the back legs, four on the front—if you looked closely. They were stretched far apart like the fingers of a piano player reaching for a two-octave chord. On the ends they had little suction knobs which enabled him to walk confidently against the glass, up and down or sideways, or just to cling there by three feet, holding the fourth out in the air as if testing whether it was raining.

He did not seem to be trying to catch any of the moths

that brushed past him. One of the larger ones even bumped him enough to loosen his grip, but he retrieved himself, undamaged, at the window ledge and returned to the glass. He may have snatched one moth, a dark-winged one the size of a navy bean. It disappeared suddenly after having been near him, and I thought I had not until then noticed the sharp angle of bulge on the left side of his pale, putty-colored body. The thin, opaque membrane of his underside pulsated rapidly and constantly. His mouth, wide and flat like a frog's, remained tight shut. His headlight eyes bulged from the topside corners of his face. His top body color was a frog's green-brown, slightly cobbled in texture, and gleamed wetly in the light.

A tree toad is said to be able to change color when moved from one background to another, but it takes at least two hours and considerable effort on his part to accomplish this. Thinking he would make a charming pet if provided with proper food and shelter, I went outside and captured him with a glass jar and cardboard, and brought him into the living room. But he was not at ease there. He went into a panic, walked about distressedly, so I returned him to his outdoor place and he immediately resumed his calm. He walked on the wet windowpane. At bedtime I was reluctant to turn out the light, expecting him to be gone in the morning. He was, and has not come back. But some rainy autumn afternoon I will hear him singing his song like a long dotted line, and I will go out and find him clinging to the wet bark of a maple tree.

OCTOBER

[OCTOBER]

❧ ❧ ❧ ❧ OCTOBER inherits summer's hand-me-downs: the last of the ironweed, its purple silken tatters turning brown, and the tiny starry white asters tumbling untidily on the ground like children rolling with laughter; stiff, drying black-eyed Susans whose dark eyes gleamed from July's roadsides; coneflowers with deep yellow petals surrounding brown pincushion centers from which bumblebees still are sipping late honey. The assignment of yellow is taken up now by thin-leafed wild sunflowers and artichokes.

Queen Anne's lace is now an heirloom, a faded wedding veil, and the mullein's tall stalk holds a few late, yellow flowers like drops melted and run down the side of a candle.

The pods of autumn are as richly expressive as summer's flowers. The quiet perfection of a long flat honey-locust pod, wine-colored and polished as the leather of an alligator-skin purse, is pleasant to contemplate; or the pungent roughness of a green walnut, for what else is a walnut except a pod containing one fruit, whose passion is to free itself from the pod and become another tree, producing a million more pods?

In the infinite variety of design, color, size, and texture, the expressed intent is always the same, to survive and perpetuate. The comfort afforded mankind by this expressed intent is, indubitably, the basis of our pleasure in winter bouquets of bittersweet, pods, or weedstalks.

This week the pods were ripening on the old catalpa trees just this side of the covered bridge over Beanblossom. The old, bent trees have that look of sparseness and authority one observes in Japanese-garden trees. Their rough-textured, heart-shaped leaves are yellowing and dropping now. The green and brown of them is blended softly like the colors of an old woven wool coverlet. The long pods hang darkening and drying, gracefully curved like the arc of a steer's horn. They are deeply ridged and shine as if they had been polished. These stark, carven, beautiful pods, suggestive of artifacts dug up from some ancient, long-buried ruin, come in autumn from summer's bubblelike fragrant white and yellow catalpa bloom, loved by bees.

Someone had asked a question

❦ In the afternoon I went up the hill to the field where Dick was disking the ground so he could sow wheat there next week. "Come up and see how nice the new pasture is," he had insisted.

The clovers and grasses planted in spring have made a deep sod, green and succulent. "Here's Ladino, here's red clover, here's alsike, here's alfalfa." Oats, planted with these for a nurse crop, were mowed long ago. The pasture was mowed late this fall and now, free of dead weeds and brush, it rolls cleanly to the very edge of the woods and there is stopped by masses of trees in the slowly deepening, magnificent colors of autumn.

"Come out a little farther," Dick kept saying. We walked across the good, soil-holding sod. "I could just lie down and roll in it," he exclaimed finally, in the fullness of his rejoicing.

In the evening, having gathered up the young pullets that stubbornly will not accept the laying house and nightly return to the brooder house where they spent their chickhood, we stopped to admire the swamp maple in the springlot. It is an old tree; its top is as big as the farmhouse. In spring it was bright red satin. This fall it has colored earlier and more richly than the adjacent soft maples. Standing high against the bright blue evening sky, with the western sunlight full upon it, the swamp maple was genuinely spectacular. If there had been nothing else of beauty or goodness on the whole hundred and thirty acres, that one tree would have been enough.

After the family had gone to bed, I went out and sat on

the back steps, from where I could see the looming shape of the swamp maple. I was restless, troubled by something I had read early in the morning. "What is the meaning of life? Man has lost the answer," a noted theologian had said.

All day, going over the farm, I had carried the unquiet feeling that I, personally, must make some reply.

Now in the rich, quiet, autumn night I began to think I had the beginnings of a reply, picked up from the clovers and grasses, the tree, and from the very soil itself.

The trees were only shadows, their brilliant color blotted out by night. Above them, the sky was a dark bowl into which someone had swept all the crumbs from a silver loaf. Night insects sang, not with their earlier, noisy eagerness. Their voices had a sleepy, faltering sound, from which one little voice rose above all the rest. It was the cricket; it sang in a tiny, tinkling sound, like a small bell being gently shaken in a sleepy hand.

The air was not cold. I sat on the stone step thinking how rich in quiet, small, accumulative ways life is, and how from each repeated year—never identical, but always recognizably similar—the pattern emerges more clear and exciting.

"Man has lost the answer."

But did he ever really have the answer? Is not his whole existence a reaching out to touch and understand the meaning of life and of himself?

Is not that the goal toward which man is slowly, unsteadily, wastefully striving? And as he goes, man has to adjust himself to the environment which, in his curiously destructive and sometimes prescient way, he creates.

Like the swamp maple, man must go through his various seasons, his burgeoning and declining. He is always on the verge of destroying everything and himself, and simultaneously always on the verge of understanding everything and himself. The two verges rise like opposing hilltops out of a deep hollow. Through many hollows, up many hills, man goes up and down, up and down.

The night wind ran softly down from the new pasture on the hill, and I remembered again the firm, resilient feeling of the new sod up there.

Our greatest achievements are only the beginnings of man's answer to the question: what is the meaning of life? Man has never really had the answer, but eternity is the time alloted him for finding it, and eternity is a long time. Much longer than the mind of man can imagine.

He has hints of his future, from the vast unexplored areas of his subconscious mind. The present intelligent area of man's mind compares to its potential as a tiny cleared hand's width compares to a whole new continent.

This was a reply. It was a fumbling, incomplete, and elusive one, but it came from the whole farm, and it was enough to let me sleep.

A LAYING HEN has her special way of announcing her resignation "as of today." She lays a half-sized egg that has no yolk. When a banty hen resigns, leaving this half-sized announcement, it is a masterpiece of understatement. More round than her usual small egg, it is no larger than a robin's egg, but concisely worded and easy to read.

THE DAY was fine; blue air and coloring leaf, happy song of bird and bug, water running gaily in side ditches; the kind of day when everything seems possible, including the sale and purchase of expensive new farm machinery.

We were going to Charlie Frye's farm where Russell Fyffe, our neighbor who sells farm implements, was going to demonstrate the new two-row corn-picker, which is mounted on a tricycle tractor.

I took along a book, expecting to have to wait, but instead rode on the corn wagon. "There, you can sit on that," said the young mechanic they called Fearnot, and gallantly spread out a square of machinery-cleaning cloth on the wagon bed at the back. He picked up one ear of the heavy blunt-tipped yellow corn and twisted it in his hands. "That's good horse corn," he said. He raises rabbits.

Russell let Dick drive the tractor. Russell sat beside him. Mr. Frye just stood in the field and watched the picker go down the rows. Mr. Frye has an old tractor, which, for reasons best known to himself, he calls "Old Satan."

Anyone could have walked as fast as the picker went. Champion pickers drive almost universally in first gear, but farmers like to show off by driving in third, which is faster.

From two rows in front of the picker's snout the corn was snatched off quickly, clean of husk, and carried up the narrow red elevator from which it gushed like a yellow fountain into the wagon bed.

"We got fifteen bushels on the first round," said Fearnot happily. The first round, two cornrows on all four sides of that field, would have been a quarter of a mile. A person who has shucked corn by hand would realize that this is fast.

"Lift the nose well up," warned Fearnot at the end of the row.

The picker's nose resembles the cowcatcher of an old-fashioned locomotive. It is accurately balanced so that the picker puts no more weight on the front tires than when the tractor stands naked and ready for a different farming implement to be attached.

At the end of that round Dick shut off the power and Russell began to remove bits of shuck and fodder from the stilled rollers. "Any piece of machinery is dangerous if a man is darn' fool enough to get careless," he said. The sides of the picker were generously posted with notices about shutting off power, keeping shields in place, and the easy terms by which the picker was available.

"This is the golden era of farming," I said as we drove home, past lines of gorgeously colored trees. We were traveling a stoned road along which box elders displayed their flat draperylike pods, honey locusts offered dark leather hangings, greenbriers had climbed into the tops of sassafras and redbud, and dark blue gentians stood at the grassy edge.

"There's all this," I marveled, "and power machinery, too." That is, if you can afford the machinery necessary to work enough land to pay for the machinery.

AT BREAKFAST Sunday we were talking about cowbirds, and by one of those coincidences that happen only in real life, and therefore can't be used in books, Dick came hurrying down from the barn later to ask: "Would you like to see some cowbirds?"

I set the glass on the table with the dishtowel still in it, and to hurry me still more he put his arm through mine and ran. Our going aroused the curiosity of Rose, who leaped up and ran ahead of us, lest she miss something. If cowbirds could laugh, these would have reason to.

We stopped inside the tool shed and peered through the cracks of the wall. In the west clover field were literally hundreds of birds feeding. The females were dull-colored, the green-black males glittered in the sunlight. As they ate, they uttered sharp, squeaky-hinge cries, overlapping into a sustained screech.

Dick said: "They're banding up, ready to fly south." They moved steadily forward in the leapfrog way peculiar to them. The ones at the rear rose up, high enough to clear the ones in front, and then settled down ahead of them again. Moving in this continuous way, they had the appearance of a black mass rolling uphill. Presently they all rose, no higher than the height of a corner cupboard, and flew away, like a scattering of dry leaves in a late autumn wind.

They are called cowbirds because they often feed where cattle are grazing. The cattle disturb grasshoppers and other flying insects, which simplifies meal-catching for the cowbirds. It would be an example of symbiosis, except that the cattle get no reward for their neighborly services; even so, they fare better than the grasshoppers.

❦ FOR WEEKS farm ponds, creeks, and water holes have been drying up. It has been hard going for fish and frogs and farmers.

Farmers who can wring out the necessary fourteen dollars an hour are taking advantage of the drought to have ponds "dozed out," small ponds deepened, new ones dug.

Yesterday morning the bulldozer began at 6 o'clock and by early afternoon had deepened the dry pond in Carr's side field and dug another one, sixteen feet deep at the center, back of the barn. Even at that depth the earth was alarmingly dry; the caterpillar's steel tread left only an occasional damp slide-print. It's been only by the grace of dew, which rises every night in extremely dry weather, that shallow-rooted plants have been able to survive. In years to come a tree cutter, reading the tree's rings, will find this recorded as a cruelly dry summer.

"I measured the water in the house cistern last week," said Carr, who can usually laugh over anything if it helps. "It was only three feet deep, and I've been afraid to measure it since."

Yesterday evening at supper Joe said: "His new pond

cost just about the price of one good steer, and the steer couldn't have gone for a better purpose."

❦ FOR THREE weeks country roads had been dusty but magnificent. Hickory nuts were falling; papaws had gone. Persimmons, even those that stay puckery until frosted, were falling and edible. Bittersweet had opened its round globes, disclosing the combination of red and orange that nothing else would dare use. Its inconspicuous leaves were dropping from stems that ended in straggly curl, like a little girl's hair that needs redoing in pin curls. Up on the hill, sumac leaves were strips of bright red patent leather.

Nights grew chilly. It was pure luxury to have a little stack of boards broken up for an evening's blaze in the fireplace. By 4 o'clock in the afternoon the hunter's moon came up, full and heavy, like a cow coming in early, with dripping udder, from pasture. It pushed upward, shoving the sun back behind the hill. By bedtime it stood at the very top of the sky, casting its burning light straight down; and under it the thirsty earth waited. When I went outside for firewood, I could almost hear the thirsty whispers rising up from all the insects, roots, and little wild animals over the farm.

When the moon wanes, farmers say, the rain will come. At midmorning Saturday Dick said: "You know I've been having to dig out the water hole every morning and I've been telling everyone the spring is dry." He accepted a doughnut and a cup of coffee. "Well, this morning I went down there and the spring had started running again. So now I know it's going to rain soon. I don't know why it is, but after there's been a long dry spell, the creeks and springs start flowing again just ahead of the rain."

He was right. Monday evening the long-yearned-for rain began. At first it was hardly more than a heavy fog, gathering in big drops and dripping from the trees. It brought out an exciting clean smell from the dry leaves. All night it gathered strength. The big drops falling from tilted leaves came down on the ground with a sound like that of small wild creatures prowling outside. By morning it had become a heavy rain. When Dick came in with the milk, the sleeves of his jacket were dark with rain.

"It's a nice morning, isn't it?" remarked Joe, ready to dart out as soon as he saw the school bus.

"Oh, it's a wonderful day to be feeding cattle and working around," exclaimed Dick. It was perfect weather: it was raining.

A PETITION was being circulated to get our road blacktopped. Signers had to be freeholders, living along the Maple Grove Road and using it frequently.

From here the petition was to go to Russell's farm, which adjoins ours, so I decided to walk through the woods instead of around the road. Rose went along. In the woods there is much for a dog to read, and Rose reads well.

The narrow, tractor-made road through the woods was like an unrolled scroll inscribed with many signatures. As she ran eagerly ahead of me, Rose added her own. Other freeholders had signed ahead of her, some carefully, some in a hurried scrawl. There were the raccoon's long print, like a baby's foot even to the little toes at the end; the terse, close-set signature of a squirrel with pencil thickness of bony fingers pressed close together; the cat's little clusters of round toes like a cloverleaf roll in a muffin pan; the larger, similar clusters made by a dog on the run. There were the flat, embossed mark of a ground hog; the long, illiterate scratch of a rabbit pushing himself forward with strong hind legs. There were many small forked scrawls of birds.

I looked for the signature of the fox who lives—we know—in the sunken spot where long ago some stone was quarried out of the woods; but the fox had walked warily, reluctant to put his signature on anything, even after reading the fine print. All the freeholders had passed there ahead of Rose and me, hurrying to their private adventures. For these little citizens, who cannot carry paper and pencil, dust is the perfect writing material in summer, and snow in winter.

BILL was to leave on Tuesday for the Army. He joined the church on Sunday morning and at his request was baptized that afternoon in the creek just beyond the Moll de Lapp bridge.

The youngest of seven children, Bill is tall, darkly brown-eyed, quiet in the way a person is quiet after hearty laughter. He had been studying nuclear physics at the university when Army training interrupted, and planned to resume studying afterward.

A week earlier he had been given a hilarious and affectionate farewell by the young people of the community. It began with a hay ride over country roads in a Jeep-pulled hay wagon, and ended with a bonfire and wiener roast.

On Sunday afternoon the same young people, with parents and friends, assembled soberly to witness Bill's baptizing.

The long, wooden bridge is set high above the creek, because in spring the swollen floodwaters overflow into adjacent cornfields. Once in the horse and buggy days, my neighbor Fanny Dunning remembers, a country preacher tried to ford the creek "and the horse and buggy were both drowned."

The bridge has open sides; low bannisters; a loose, plank floor.

Just beyond the bridge the creek deepens, making a place where farm boys swim in summer, concealed by the vine-tangled lofty tops of trees that meet above the water. This was the place chosen for the baptism.

The autumn afternoon was cool. On the shadowed surface of the slow-moving water a handful of leaves drifted like abandoned small boats. It takes some deep feeling, I thought, to cause a young man to want to be baptized here in this austerely simple place.

During the customary hymns and prayers Bill and his parents stood a short distance apart from us. The minister held out his hands and Bill walked into the dark water to take them. It was an impressive, reverent ceremony; the smallest children watched in perfect silence.

We were in a world apart from our everyday one, but none of us realized how deeply withdrawn we had been until abruptly a car passed over the bridge, its passengers unaware of the scene below. The loose planks rattled briefly, bringing the outside world into our hidden one. It was a reminder that there is no real withdrawal, that the everyday world is always there, to be confronted, and that there is strength with which to confront it. That was Bill's farewell gift to us.

❦ AT the church-house field where Dick was mowing, blackberry briers were a gorgeous red and goldenrod gleamed behind the cutter bar. "I always leave the five-leafed ivy just for the sake of this one short season," he said, and pointed to the woods where three tall trees were strung with heavy ropes of the vine, now a dark red.

"Looks like Christmas decorations," he continued. "If you stop over there by that old stone fence post, you'll find a bird's nest I saved for you." I had already found it and had it in my hand. "Thank you," I said.

"I wish when you go back to the house you'd look in the almanac and see what the sign was last Thursday when Warren and I worked on the hogs. I never had a bunch do better."

It was Libra, Balance, Kidneys. "Plant root crops.

Fair for grain. Next best after Cancer for planting." Not a word about castrating hogs. Not a word against it, either. It's on such delicate balance of opinion that a farmer anchors his belief in signs, having first made it clear that he considers it all pure superstition.

❦ INDIANA's autumn leaf coloring, like the Grand Canyon, is so magnificent and spectacular that the only way you can describe it, or even believe it, is to understate it.

Let it be said, therefore, that in a hillside around which the Maple Grove Road climbs on its way into town, there is a young beech tree that has authority to speak for all beech trees. Its coloring bronzy-yellow leaves shine with a metallic glitter. Enough of them have dropped to make the ground glitter also, around the tree, and yet the tree is not diminished of its glory. Anyone who has seen it now will remember its light all winter.

Throw in sweet gum and ash for purple; sassafras for sentimental pink and deeper red; sumac for a glossy bright red; luminous yellow of poplar and papaw and hickory and maple. Add the kaleidoscopic pink, yellow, rose, and red of maples; the quiet, mighty old-hymn brown of oak; the warm, live brown of tall white sycamore; the dark, melodious red of wild blackberry's rough leaves; the warted yellowing brown of hackberry; the dangerous brilliance, even, of reddening poison ivy; the sharp, durable, bright green of greenbrier. And there are yet others. In autumn not an acre, hardly a foot, of rural Indiana is lacking in splendor. It is gaudy, breath-taking, unbelievable—it is Indiana in autumn dress.

❦ Hunting wild grapes for jelly, I went up yesterday to the southwest pasture. As I crawled under the barbed-wire fence, a lean, brown hound got up from in front of the low stone fence at the church and came down to walk with me. He had a long, friendly face and a companionable silence, and I was glad he had come. We stopped to pick up some beer cans from the road and toss them into a sinkhole, and when I stepped back my wool coat was besieged by burrs.

All kinds are eager to travel now. On my coat were the round, hedgehog kind, the size of BB shot. There were also those that grow like a bundle of two-pronged flat splinters in fagots at the center of a yellow flower. Farther on were the dark brown, oval cockles, with wirelike, incurling prongs. There were some sticktights, flat and triangular, like glazier's points.

We stopped at the apple tree in the fencerow. The ground there was strewn with spoiling apples that had been sour and red. Now they spiced the air with a winy smell. The apple tree is tattooed in neatly dotted lines by the searchings of a woodpecker. It is in its prime; its wide-spreading branches touch the ground upon which it has cast millions of seeds, only one of which has survived to make another apple tree.

We walked across the late-mowed field and the brown

dog licked my hand when I reached down to pick some five-leaf clovers I saw.

The field was a signed statement of nature's confidence of survival. New blackberry briers and Queen Anne's lace, sprouted after the late mowing, are eight inches high. New milkweed is taller. The older, mature relatives of these, out in the unmowed roadside, have long since opened their pods and released their silken parachutes, each bearing its one seed away on an assignment of posterity. But the young milkweeds are as confident as if they had plenty of time to ripen. What a marvelous, heartening spirit, compounded of hope and ignorance! Perhaps the world owes more to ignorance than we realize.

Across the creek I noticed sadly that the big elm is finally dead. Its look of desolation was slightly alleviated by the knotted mass of grapevines growing in its top, darkly dotted with small, not very juicy grapes. A tufted titmouse hopped among the branches, feasting and exclaiming to itself. I broke off some low-hanging clusters and turned back from the field. At the gate my companion left me, returning to the stone fence to wait confidently for his owner. I hoped his faith was going to be justified, and went on down the hill, noting with pleasure that the sumac berries were deep red velvet, and if they were used to dye shoe leather—what beautiful shoes these should be!

❦ EVEN FROM a distance it was obvious that the white cow just wanted to be alone. It showed in the way she walked, not fast but not placidly either, and in the way she stopped now and then to pull off a bite of grass, never really putting her heart into it. It showed in the restrained impatience with which she switched her tail. A cow's tail is one of her most eloquent features.

Walking beside her at identical pace, their heads just even with her flanks, one on each side of her, were the two calves she has nurtured this summer, her own and a foster daughter. They followed her with cocklebur tenacity. When she stopped, they stopped. When she walked, they walked. If she turned to right or left or reversed her direction, they did the same. It was enough to drive a mother crazy.

The white cow, a large-boned, polled Shorthorn is dry now and due to calve again late this month. When she is fresh, she provides a generous lunch and has done well by the two calves. They have outgrown milk and actually live on grass, but it had not occurred to them that the white cow had outgrown them. A mother dog would have growled and bared her teeth; a cat would have scuffed the kittens out of the way; a sow would have stood up and run, squealing. All the white cow could do was to walk, switching her tail.

At suppertime Dick said: "I separated the white cow from the calves. You may hear 'em bawling for a while. It's nothing to worry about." I went out to offer my felicitations to the white cow. In her new freedom she had lain down by the big pond and was chewing her cud in the bliss of solitude. The calves did bawl awhile; then they recovered from the astonishment of their independence and, discovering they liked the taste of it, they quit bawling.

❦ "I'll tell you how you can tell the farm women from the others at a meeting," said Dick after he had been to a meeting of the County Extension Committee. "You look around, and you can tell 'em by the scratches and bruises on their shins. You can see 'em, you know, right through their stockings."

I know.

❦ It rained last night—"away along in the night," as farmers say—without thunder, but with intermittent, quickly extinguished flashes of lightning as if someone were trying out a new flashlight.

The noise of rain was soft, swept-sounding, like newspapers being blown across a bare floor. I would not have awakened but for the noise of the kitchen door being opened by Rose. She learned long ago that if she puts her front paws on the doorknob and lets her body fall sideways, her weight will turn the knob and open the door.

She came in because she is afraid of storms and wanted someone to get out of bed, take her up to the barn, and open the door so she could go in where the cattle are.

In ordinary weather Rose is jealous of her position of authority over cattle and hogs. But in a storm she sheds this blessing of rank gladly in exchange for companionship. She likes the dark security of the barn where there are animal companions, warm, fragrant, breathing, and unafraid. Authority is sometimes a lonely possession.

❦ FOG FILLED all the space between earth and sky. When the sun came up, the fog turned it to a silver ball the size of a child's geography globe. Fog made fragile things spectacular and solid things obscure.

"Come out!" exclaimed Dick. "You must see this wonderful spiderweb in the mock-orange bush." The web hung vertically like a kite. It was fourteen inches across, twenty up and down. From the center, twenty-seven delicate threads reached out to the outer threads that anchored the web to the mock-orange bush and kept it taut. The pattern was made more intricate and unusual by little irregular crisscrossing threads, filling in spaces and joining the twenty-seven basic ones. In the whole web there was enough thread to have stitched up a pair of pillowcases, and all this had been spun out in the foggy night by one spider, from her own body.

Other spiders had built similar webs nearby, none so beautiful, although one in the seedling peach tree was twice as large.

Fog made the webs spectacular. Every delicate thread was strung with silver beads of fog, and, wherever two threads met, a larger bead gleamed. The silver sunlight brought a silver gleam, but no glitter, from the beads.

In the still air the web trembled, but the beads did not shake off nor run together. Even on vertical threads they remained entirely separate.

Webs made by other spiders lay in the grass or hung from asparagus stalks, from raspberry canes and tall weeds in the garden. They were clumsily constructed, heavily reinforced by crisscrossed braces, and some looked like a handful of cotton candy dropped in the grass. In their whiteness, they applauded the bead-strung, delicate web that hung, intact and trembling and splendid, from the mock-orange bush.

❦ FENCE BUILDING, like the regrowing of skin, goes on all through a farmer's life.

Fundamentally a man doesn't mind it, but he doesn't like to do it alone. That's why fences along the road are always in better shape than those back on the farm where nobody is likely to stop and interrupt a fence builder by visiting. Fencing brings out the social nature of a farmer, makes him realize that humankind needs neighbors more than most creatures need them, and needs them longer.

The flow of genial conversation follows the line of fence, supported by solid posts and jokes and barbed with wit and wire and exchange of opinion.

A wife is a good fencing companion. She is naturally susceptible to a farmer's "I'm so lonely" manner and his assumption that of course she will be glad to get away from the dreary, lonely monotony of housework or whatever else she has cleared time to do. She is eager to help set fence posts and stretch barbed wire on a hilltop where there is a splendid view to admire and a fascinating, flattering man to talk to. At least this is the impression Dick succeeds in giving when he comes to kiss me good-by before starting out to build fence.

(The USDA has no statistics available about the per cent of farmers who kiss their wives good-by before leaving for the field. It could run from fifteen to eighty-five per cent.)

Fence building is easier than it used to be when the children were too little to do anything except go along and play in the woods, which offers in season all that a woods can of wild flowers, mushrooms, animal tracks, bird song, wild raspberries, blackberries, mulberries, May apples, papaws, walnuts, sassafras, trees to climb, holes to explore.

Now we have a giant steel screw which fastens to the back of the tractor and digs a good posthole so quickly that a man can make money custom-digging, at fifteen cents a posthole, and not even get off the tractor seat.

Before we got the power digger, it was this way: Preparing to go, Dick handed me the fence stretcher, a complicated assembly of wood and iron and dangling chains, and the bucket of fence "steeples." He hung the claw hammer in the tab placed for that purpose on his overalls, picked up a three-foot iron claw bar and the hand posthole digger. Together we carried a roll of woven wire to the tractor. With this, a little lunch of crackers and oranges and a can of drinking water, and the two children and Rose, we were ready to start. Joe drove the tractor (a farm boy can drive a tractor from the age of seven, even when he has to stand up to put his foot on the clutch pedal); Carol sat on my lap, holding the lunch.

At the hilltop a roll of barbed wire was already waiting. First we set the corner post. The hole for a corner post is three to four feet deep; for line posts, a little less. The hand-powered posthole digger is, basically, two long wooden handles fastened together like scissors. On the lower end of each handle is a curved steel scoop, pointed at the end. When these scoops are struck forcibly into the ground, they bite out a four-inch depth of earth, which is then removed by pulling the handles as far apart as possible and lifting. The loose earth is laid respectfully aside, none scattered, because even with the added bulk of the post, the removed earth is never enough to refill the hole solidly.

Next we stretched the guideline, a length of barbed wire. There is probably nothing that makes the human hand more acutely aware of its tearable humanness than a string of barbed wire can.

The best posts are black locust, or red cedar. Hedge apple is durable but hard to drive "steeples" into. "It'll last a hundred years and then turn to stone," Uncle Bent Stanger used to say.

With all the posts in place, Dick likes to look down the line critically. They must stand up in a smartly military way so that when you look you see only one post. We unrolled the woven wire in homage at their feet and attached the fence stretcher to the end with bolts, nuts, and pulleys.

Although I operated the stretcher, I never understood it. I simply did what Dick said to do, and the fence came out taut and proud, ready to be "steepled" to the posts, and was stockproof.

At this point, in the days of the hand-powered posthole digger, it was time to stop and do the evening chores and get supper. We retrieved the children from the woods, got on the tractor, and returned to the farmhouse. There was something genuinely appealing about the old way of building fence. It could have been the farmer, of course; but I must admit that fence building is easier now that Joe takes my place and part of Dick's authority and the mechanical digger has replaced the hand-powered, scissors type.

❦ It is doubtful that any insect has solved the complexities of personal nourishment and storage better than the vinegar gnat has.

If you need some for research on this point, you need only lay a bitten apple on the kitchen table, or a few grapes pulled from their stems, or a tomato weeping over its wasted life. Vinegar gnats will suddenly be there. From where, goodness knows. They arrive quietly, requiring no more space than wind requires. Their sense of news must be prodigious.

They neither bite nor sting, buzz, chew up clothing, or build webs in which to trap other insects. They are able to survive deep cold; if a gnat accidentally gets shut up inside the refrigerator, he is able to resume his activities when the door is opened, unless in the meantime he has fallen into an uncovered bowl of fruit or milk.

Gnats never seem to fly very fast, but they fly elusively, as every housekeeper knows who has tried to clap her hands together into a gnat sandwich.

In the scientific laboratory, where it is known as the fruit fly or, specifically, the Drosophila melanogaster, the vinegar gnat has made a valuable contribution to the study of genetics. Because the fruit fly's life cycle is extremely short —within less than two weeks it goes from tiny, elongated white eggs to longer, pointed white larvae; to blunter, fatter, pale brown pupae; to six-legged, winged, silent-flying adult flies—geneticists have been able to study many generations of these gnats within a short time. Through his work with fruit flies, geneticist Thomas Hunt Morgan was able to demonstrate his theory of genes as the determining factors in heredity. The study of American fruit flies has given United States geneticists most of their present knowledge of the structure of chromosomes and the behavior of genes.

Decidedly, with such an outstanding scientific career possible, the vinegar gnat is wasting its time in the farm kitchen.

❦ FOR SOME PEOPLE autumn is summed up in the magnificent drama of colored leaf. For me the most poignant, almost unbearably dramatic, thing about it is the passage of wild geese, southbound in a chilly autumn evening, over the farm. There was one time in particular, when the children were little:

We had gathered walnuts from under the bare trees on the hilltop one chilly evening after school, and were bringing them down, three bushels of them, in the children's little red wagon. By great effort of pulling over the rough sod, we had got as far as the fence. Joe exclaimed suddenly: "Look! Wild geese!" He was pointing to the northeast, above the barn, and we all stopped, thrilled, to watch.

They came swiftly, a long, dark line undulating against the blue evening sky, like a dark ribbon tossed into water. They were high above us, flying diagonally across the clover field. We could see the leaders at the head of the inverted V, and we could hear the weariness in the lonely, beautiful, wild crying. We watched until they had disappeared entirely, and it was as if some part of us had gone with them.

As we turned back to the wagon in the darkening evening, I was aware that some important truth had been summed up in the flight of the dark, vanished birds. It was stated in the cold, clean fragrant air, in the crackling of dry leaves under our feet, and in the blended colors of far-off hills, where already many emptied trees were only bare, dark outlines. It was a statement of acceptance, but not of resignation. I was reluctant to start on. There was something there that was immensely lonely and vulnerable, and yet invincible; and I wanted the children to sense this poignancy in which transience and permanence exist side by side, in the same thing.

The wagon skidded on the brusied, succulent clover as the children gave it an impatient tug, wanting to go on.

I wished to tell them: "Remember this—feel it deeply —there is something vital here and you will understand it later."

But they were cold and hungry and wanted to be done with the walnuts. We went on down the hill, to the house. Carol picked out the black nuts whose hulls cattle had trodden off, and Joe spread them on the brooder-house roof, where rain would wash them and sun would dry them. I poured the unhulled ones into the driveway; there incoming tires would mash off the hulls. By that time the earth was fully dark and we hurried into the kitchen, to warmth and light and a supper of bread and milk, roast pork, and apple pie. A good finale to a symphony of wild geese and autumn.

❦ EVERY YEAR, late in autumn, this happens.

There comes a day of bright sunlight with a minimum of wind, and suddenly the farm fields are hung with a floating, shimmering, silvery glint. From dry weedstalks and the pointed tips of countless grass-blades to which they are fastened, a myriad of exquisitely fine, silken spider threads float, all in one direction. You cannot see where they end. You cannot even see the end. If you come close enough to try to follow one thread to its end, it disappears. It is visible only as the sunlight touches it, and is actually more clearly visible from a short distance away than from close up.

If you try to examine one thread through a microscope,

the microscope shows only nothingness magnified. At one place that seemed to be a knot in the thread I looked through a microscope and saw only a tiny foam-fleck. The long thread of silvery shimmer is the only evidence that anything is there. Whether these are old threads released by spiders for traveling, or the unwanted remnants of web material thrown away from their spinnerets, or parachutes bearing spider eggs, is uncertain. The phenomenon is as seasonal as the sudden, inexplicable mushrooms of spring, and as perennial as autumn.

NOW THE COLOR in hillsides has burned down to embers, is darkening and dying out. The several entirely bare trees that stand grayly among those with lingering color create an illusion of small ribbons of smoke rising up from a nearly spent bonfire.

This is the time of year when you may be able to pick a last "mess" of greens from newly sprouted wild mustard in sheltered places around the barnlot; when old, weather-beaten barns hunch their shoulders like old men gratefully soaking up the warmth of late autumn sunlight; when hickory nuts have fallen, bone-white and not quite dry, as their thick outside hulls split into quarters and fall away.

It is the time of year when, as you sit by the window reading, your page is likely to be darkened by small, hurrying shadows; and you look out of the window to see whether it was a flock of small summer birds migrating south or a flock of hurrying leaves with the frosty wind nipping their heels.

"Did you know broom sedge has a silky down at this time of year?" Dick asked, handing me the evidence. "It's a pity the birds aren't around now to use it in their nests."

THERE COMES a time finally when it cannot be put off any longer. The radio warns of a killing frost coming that night, and you must say good-by to the garden. You dread it, as you dread saying good-by to any good friend; but the garden waits with its last gifts, and you must go out with a bushel basket or big buckets to receive them.

The late afternoon sun gives off a grudging warmth. You wear a coat. Under your feet the dry leaves make a crisp crackle and at the garden a tardy cricket is singing. His voice is old and rusted and he will not try to leap out of your way as you walk past.

In the garden there are still a few ripe tomatoes. From some the skin can be peeled off in a thick, mealy layer. Others, large and shiny, are extra-fine. They are, of course, the ones chosen to be pecked into by the determined little gold-speckled hen that finally discovered a way to get into the garden. There are many firm, shining, green tomatoes on the old, tired vines that have now sprawled out on the ground as if to take a long, earned nap, cushioned by thrifty foxtail grass.

Gathered now and sheltered in the woodshed, where

already the green-picked Kieffer pears are yellowing, are the piles of late-ripened bean plants, waiting until somebody gets time to shell out the dry beans. Looking carefully, you can find enough more tender green pods and shellouts to make one last delicious cooking of fresh beans.

There will be a few foolish late onions, too, with tops as green as if they were starting a new summer right now, although the sets on dry stalks of potato onions have already taken root.

Take also the small beets, carrots, and turnips you missed on an earlier harvesting. Pull off the last tender okra pods from the bare, stiff stalks. It is time now, too, to bring in all the squashes, for frost will fall alike on the ripe and the unripe. Bring the few remaining dark-green acorns; the long green- and white-striped crooknecks; and the dark, rough Hubbards. Ready or not, here they must come. Bring the small, dry ears of sweet corn to hang in the trees for birds. Bring any remaining sunflower seed cushions.

All this you can set on the back porch to be sorted, being careful not to bruise the green tomatoes, for with care you can have ripe tomatoes until well after Thanksgiving.

Then you take a paring knife and go back to the garden to get the last of the happy-go-lucky zinnias, the furry red coxcombs, the orange and yellow marigolds with the old-fashioned pungent smell, the delicate but faithful pinks, and the rose cosmos. This is the final affectionate farewell of your garden, from which you have received so much. In return, you pause long enough to make your perennial vow to tend it more faithfully next year. Now the garden can settle down happily to a long, necessary, and earned winter's rest.

❦ THE FROST gave a sample of its intentions, but there has not yet been a really hard, killing freeze. Scarlet sage still lifts its red torches above vigorous leaves; nasturtium makes new buds.

But insects have felt the coldness in their tiny "bones." This week caterpillars crawled to the sudden shelter of anything that was handy. You could take your own count of them, dark ones and light ones, to determine whether they prophesied a cold or a mild winter. Grasshoppers, katydids, crickets—all the singing autumn insects had subsided this week. If you saw one on weedstalk or window sill, it was stiff and sluggish. Katydids retained their brightness even in these stiff postures, but grasshoppers had a frostbitten blackness; and in the cold nights of this week, finally, the cricket was still.

NOVEMBER

[NOVEMBER]

♣ ♣ ♣ ♣ I'LL have wood for the fireplace when your club meets here if I have to tear boards off of the barn," declared Dick yesterday. It was as truly a declaration of love as an armload of roses would have been. Only love could induce a farmer to tear boards from barn walls, especially when the lower half of the wall is of concrete blocks and to get any boards a man would have to climb a ladder or dangle himself from the thin edge of a galvanized roof.

♣ ALL NIGHT the wind ran its fingers through the maple trees and by morning had tossed down a luxurious accumulation of dry branches, excellent kindling for the fireplace or the little drum stove in the back room. Dry maple burns with a quick snapping, throwing out sparks.

On the porch this morning, also, was a stack of freshly sawed hackberry, its ends showing the lines of growth and still lightly dusted with sawdust.

Hackberry has a warty, rough bark. Close to its leaves grow the little clusters of stems called witches'-brooms. The wood burns with a bright flame and a pleasant nutlike fragrance.

In November a rick of wood on the front porch is as decorative as the bowl of bittersweet in the living room. For that matter, in these days of bottled gas and electricity, a farm woodpile, neatly ricked up and braced against the prowling, sharp-clawed wind, is pure luxury. It is beautiful in winter, topped with snow. If any is left over by summer, it is beautiful with sweet peas climbing up across its bark-strewn top.

For the fireplace philosopher it recalls memories of good friends in happy places, of long, satisfying thoughts drawn out before the glow of bright, burning-down embers. A fireplace doesn't warm a whole room, nor even a whole person, very well in deeply cold weather, but where it does warm, such as the skin over the cheeks, the backs of the legs, or the cockles of the heart, it warms generously.

POOR RICHARD's Modern Maxim: "It costs three dollars to stop at the supermarket for a loaf of bread."

Listen for the old sounds

❦ JUST AS daylight was beginning to lay hold of the farm this morning, a young rooster crowed from the henhouse and was promptly answered by roosters on all the neighboring farms. The commonplaceness of this familiar sound is pleasantly reassuring—as if, returning after a number of years to a house you remembered happily, you opened the door and found everything just as you remembered it.

It must be by some ancient, vestigial impulse that a rooster recognizes the hidden seam of morning and feels impelled to arouse the sun and the farm family to the day's work.

Even though I couldn't see the rooster, I knew just how he looked because I have watched him in daytime "crowing up a rain." He rises on his yellow toes and stretches his richly feathered neck to its full length with such intensity that his bright red wattles quiver. He shuts his eyes and pours his soul into his cry as if he had to make himself heard not only all over this neighborhood in the present morning, but also far back into the past and ahead into the future.

The rooster doesn't suspect it, but his cry is the real reason his kind has outlasted the plowhorse on the farm. Many voices, once a familiar part of farm life, have fallen silent in the wake of progress: the creaking of buggy wheels pulling up the last, long hill at the home end of the road; the gasp of the pitcher pump as it surrendered its prime against the certainty of night's freezing; the slow, starting grind of the hand-turned cream separator and its rising whine as it gained speed and finally rang a bell indicating that it had reached sufficient speed; the metallic rumble of steel-tired wagon wheels on a stony road, and the ring of a horse's shod feet. The muted, rhythmic throbbing of an electric pump has replaced the lazy, slow sigh of the windmill wheel, barely turning in a long, summer afternoon and casting its long shadow across the grass. Gone now is the shrill whistle of the steam engine on threshing day, piercing and far-reaching when the pressure was up as it arrived at the farm, and its faint feeble whisper emitted late that night when the steam had gone down and adventurous farm children slipped down to the still-warm engine and pulled the cord.

Farmers are sentimental. They give up old ways reluctantly, even while they reach out for the new improvements. In a farm world where chickens have electric lights burning all night, and balanced diets, warmed drinking water, and automatic skillets to be fried in after being machine-picked, where setting hens have vanished, farmers like the sound of permanence contained in the rooster's unchanged morning cry.

ONE OF the occupational hazards of farming is the danger of being suddenly assailed by what seems, at the moment of assailment, a revelation of great truth.

It comes, usually, from observation of small things. As I came up from the mailbox this morning, I stooped to pick up one gorgeous red maple leaf that had just broken loose from the tree and fluttered gaily down to the ground as if it had just escaped from a restraining hand.

On the ground it would disintegrate, its separate components would begin to break away from the leaf form in which, all summer, they had been united.

The two continually opposing forces of life are the effort to break away and the effort to hold together. Science now says that atoms and molecules and even smaller parts are held together unwillingly, always trying to break apart into their own individualities.

Nature says this in many ways, including the falling of colored leaves from the trees in autumn.

It is in the interplay of these two opposing forces—which seems often more grim than playful—that life advances. The little crack in the stone, pushed apart by a tiny blade of grass that will presently die, supplying foothold and nourishment for more blades of grass that will widen the crack; or the halves of a peach seed, resisting but being broken apart by frost and heat to let the tiny, white-footed root take its first sip of earth's rich juice; the skin broken on a corn grain by the upward thrust of a sharp green blade; petals falling from a ripened rose; one generation of mankind trying tenaciously to hold the old ways and beliefs together while another determinedly pulls them apart for new ones, not always better but new—this is some of the interplay.

Between these two antipodal forces, life advances.

The theory of unwilling atoms explains, perhaps, why coats slide from hangers, why things disappear and come to light only after you stop hunting them and start looking for something else; it explains why antique dishes leap out of careful hands to commit suicide on the floor. It explains runners in nylon stockings, juice streaming out of fruit pies in the oven, lint growing under beds and cobwebs on ceilings.

It is probably one of the compensations for living in an age so dedicated to science that one feels guilty for wanting to study history or French horn or pinking shears. On the other hand, with too much of science and not enough of these other things, our civilization could become like a broken zipper. The zipper key will move forward, neatly zipping up the jacket, but the line comes apart right behind it, so that when you get the key all the way up to the collar, the jacket is still wide open.

❦ SUPPER—roast beef and brown gravy, baked hominy and blackberry cobbler—was almost ready; but it was not quite dusk when we went up to the cornfield last evening. We hoped to see a flock of wild ducks that had been coming there to eat every night for the past two weeks. "John calls them 'little white-bellied ducks' and says they'll stay around as long as they can find anything to eat, unless it gets too cold," Dick said.

Where we walked, the dry stems of mowed clover made a sound like bread crusts being crunched. The cornfield was littered by an untidy fluff of stalk and blade knocked over by the cornpicker. The early-risen moon was high in the east, a pale orange-pink disk.

Dick carried a bucket of mineral supplement to pour into the cattle trough. The cows, contentedly eating the down corn, watched us with interest as we crossed the field.

Wild ducks at dusk

The sky was almost dark, and we had just about given up hope of the ducks' coming, and then suddenly they arrived. Small, dark, swift-flying, they came out of the east in a shifting pattern, uttering impatient, hoarse cries. We could barely see the white gleam of their undersides and the sharp outline of their shoulders. Behind the first group was a smaller group, trying to catch up and weave itself in; then came two smaller groups; and far at the last, two ducks, flapping and crying and trying desperately to catch up.

We ran across the field, to be there when they lighted to eat, and watched them.

Even after the ducks had disappeared into the gray distance, we stood on the high bank near the road's edge, not so much expecting them to return as to enjoy the tranquil, finished mood of the evening. Behind us, in the west, the smoke-veiled sunset thrust its pale, dissolving pinkness against the sky. Ahead of us the open fields rolled gently back to dusk-dimmed woods and hills. The darkness was punctured intermittently by lights in neighbors' houses and barns. Farmers were finishing their chores; women were getting supper.

Dick scattered a handful of supplement from the bucket to the ground, and the cattle came up eagerly.

"I'll go down and finish supper," I said.

"I'll sit here on the trough awhile and watch the cattle eat," said Dick, "and then I'll come down, too." It is one of the deepest satisfactions a farmer can have, the sense of content that comes from knowing there is plenty of food for everything—the farm livestock and the transient, unbidden, wild birds.

❦ THIS GENERATION is beset in its daily life by fears, angers, blockings of its most passionately cherished hopes, by loneliness, and by the need of solitude.

The opportunity of looking intently into some small portion of his natural environment, or the experience of being responsible for the survival of some part of it, enables a man to see his own place in the world with greater clarity, and perhaps with greater compassion.

It is only when he sees how everything is vital to the pattern and fits into it that he achieves a kind of refuge for his own buffeted spirit. It is not a hermit hiding place—the modern world has no place for a hermit—it is a kind of spiritual symbiosis. It is the beginning of the science of kindness.

❦ BY BEDTIME the moonlight was almost a violence, as if at any minute it might break out into the sound of a mighty rushing. Under the maple trees, shadows lay black and heavy; if they could have been dragged into the woodshed, they could have been chopped up for firewood. On the red cattle beyond the fence, white markings stood out like snow; and looking through the window I counted five dry leaves on the back walk. If a stranger had come up the walk just then, I could have known the color of his eyes by moonlight.

Lunar eclipse tonight

It was not only I who felt something peculiar about the intensity of this moonlight. Before midnight, Rose walked out into the open yard and expressed her feelings in a long, eerie wail toward the moon. I tapped on the windowpane and she reduced the volume, but even her whisper had the quality of a wail. The night went on; I went to sleep.

I wakened again presently, aware of abrupt, deep change. The moonlight was gone, the earth completely dark. By the clock I saw that it was after one o'clock, therefore a new day, and I remembered this was the day there was to be a partial eclipse of the moon. It would be visible over most of the world, the newspaper had said, which included the farms along the Maple Grove Road. This was the eclipse.

The moon, having risen nearly to the top of the sky, was shrunken to the size of a dinner plate. Only the extreme south rind was now moon-colored. The rest was a mixture of brown, red, and orange. The whole globe looked slippery, as if it would roll out of your arms if you tried to hold it.

I wakened the family to see the eclipse. "When the moon is in the earth's shadow," we had read in the enclyclopedia earlier, "there is a lunar eclipse. A lunar eclipse can occur only when the sun and the moon are at opposite sides of the earth."

We were seeing, on the moon, the shadow of the earth on which we stood looking at the shadow! It was a little like being on TV and seeing yourself on the TV screen at the same time, only this was awesome.

The rest of the family went back to bed, but I stayed awake to watch the eclipse at intervals. There was not much change and presently I fell asleep myself. The next time I wakened, a little after two o'clock, the eclipse was over. The moon was again whole, moon-colored, and serene, and it was Sunday morning.

HAZEL'S and Fred Dutton's house is back off the road and down a long winding slope. Solidly built about 1880 and flawlessly kept up, it is as clean inside as the inside of a pea pod. It is surrounded by a wide, tidy, fenced yard, an immaculate, generous garden. Everything grows well for Hazel. Her neighbor Floy once gave her a flimsy little start of night-blooming cereus, a plant which ordinarily blooms only about once in seven years. "I don't know whether it'll live at all," said Floy dubiously, but five years later Hazel was inviting her neighbors to come and see the six fine flowers it had produced.

In addition to all this, she is an excellent cook—things like fried rabbit, with creamy gravy to put on mounds of fluffy, steamed rice, and buttered new turnips, crisp pickles, and chocolate fudge cake.

Their five children have long since left the farm and married. For some years now Hazel has stopped helping Fred with the milking, but she still takes care of the chickens, the yard, the garden, the canning and food-freezing, and does incredible amounts of needlework.

Her whole house is a remarkable example of a woman's time lovingly spent in needlework. It's a big house: six rooms and a center hall and bathroom downstairs, two or three more rooms and a hall upstairs, and a walk-in attic that could house a whole family.

Some of her floors are covered with commercial rugs and linoleums, but even in these there are small hand-hooked or crocheted scatter rugs. The downstairs hall is covered, full length, with a hand-hooked rug forty inches wide, and smaller matching squares are at the foot of the stairway. Some of her hooked rugs have designs of flowers, some have a distinctive design patterned after the old-fashioned "hit or miss" loom-woven rag carpet.

"The ready-stencilled patterns were so confusing I just made up my own designs," she said. In these rugs she used mountains of old skirts, suits, coats, and sweaters.

For the tables over the house she has made crocheted doilies, full and ruffly and starched to cardboard stiffness. There are afghans on the sofas, white crocheted squares on the backs and arms of chairs. Yet the house does not look oppressively hand-decorated. It is pleasant and restful, a hospitable place for a neighborly visit. Hazel laughs often and gaily, and Fred laughs oftener, a great, booming laugh that is fine to hear.

"Come and see the bedspread I crocheted," invited Hazel, leading me into her guest bedroom. The geometry of white design, marked at intervals with a pink flower, made it a charming piece of handwork.

She pulled open the drawer of a fine old cherry bureau and began to show me her handmade pillowslips, taking them out gently. There were dozens of pairs, all hand-embroidered and edged with handmade lace. "I wanted them," said Hazel in a voice of quiet pleasure, but sounding practical, too, "so that if anybody was sick for a long time, the bed would look nice." A few weeks ago, late in the night, Hazel had a sudden, violent illness and, because they had no telephone, Fred had to go for the doctor five miles distant.

"And do you know," said Hazel, "the pillowslips I had on the bed that night were just plain old feedsack ones."

"I'm surprised you didn't get up and put on some of these pretty ones," I remarked.

"Never even thought of it!" she exclaimed. "I was suffering too much."

The doctor didn't notice either. He eased her discomfort, gave her medicine and advice with never a word about pillowslips.

In the long, lovely, rug-covered front hall he stopped suddenly. "Why don't you take the paint off of this beautiful old stairway and find out whether it's cherry or walnut?" he demanded.

"Because I haven't time," said Hazel. A farm woman, like everyone else, has time only for what she wants to do.

As WE WALKED up the hill to bring in the cows Dick said: "I can always tell when someone's coming up to the field, by the way the cattle throw up their heads and stare."

At the gap, where the sod was deep and luxurious as a bearskin rug, he stopped, picked up a glass coffee jar, and hung it over the top of a fence post. "This is a little game between me and the cows," he explained. "Every evening I put this can up here and every day they knock it off."

Cattle are avidly curious and quick to note any change in the farm routine.

"The calves have one habit," continued Dick, "that is sometimes funny, and sometimes not. There's a latch on the barn door and when they want in they can pull the string and get in. It's all right, except on rainy nights; they can't latch the door from outside, so it stands open and lets in rain."

"The cattle come to the barn when they see me coming," he went on, opening the long gate and swinging it far back. "I never tell them what I'm going to do, but they see me coming and just come in. Horses are different. They have as much curiosity, but they try to conceal it. They go ahead grazing as if they didn't care. The giveaway is the way their ears suddenly point forward; it's involuntary, like a person blushing. What shows a horse is alerted is the way his ears flop forward and stick up straight, instead of dangling limp or flopping at a fly."

"After you understand livestock," he added, "you just see so much in people."

ROSEMARY, the little Bantam hen, has weaned her chicks. I saw her do it.

For some time now the adventurous, half-grown chicks have been running ahead of her when they went out to the lot. Last Tuesday morning she took them out as usual, and when they had all got ahead of her, she stopped. One small white chick noticed she had paused and ran back to her. Rosemary gave him a sharp peck on the top of his head, and he ran quickly to join his brothers and sisters.

Then the little mother hen turned and ran in the opposite direction. For her the era of motherhood was over, and the next morning she went back to laying.

Now IS the time when the disparity between man's planning and nature's planning is most apparent. For nature there is a brief unused moment between essential chores; for man there is none. His schedule runs unbrokenly on; trains and newspaper deadlines to meet; milk routes to haul; school buses to get on and off of; store sales to open at the exact, appointed hour; all the artificial boundaries of his life to be kept up by him for the sake of his business.

Nature, every year, luxuriates in this brief rest, which is no more than the duration of a sixteenth note in a bar of music. But it is there, essential and definite, and it is now, and nature gives it its full value.

The yard is full of yellow maple leaves, turning a reluctant brown but not yet disintegrating. They have cooled all night, soaking up rain or dew, and in the early morning warmth they dry out, rustle again, and give off a pleasant fragrance. The summer crows gather in bare, tall trees to complain. Their cawing has the sound of being distant in time as well as in space, the nostalgic sound of something heard long ago and forgotten. It is a lonely complaint and wakens more sympathy than it did in June, when corn was sprouting.

Now through the naked trees the outlook is broadened. Distance comes in casually to partake of the farmer's horizon, like the sparrows that fly down, uninvited, to eat from the chickens' feed pans.

Like a yellow leaf blown along a country road, the school bus comes. It passes the farm mailboxes and the clumps of elderberries from which birds have now eaten most of the small, dusty berries. From long vines, now bare of leaf, hang heavy clusters of green-white, waxen berries, beautiful to see but not to touch, for this is the ivy that is no man's friend. Buckeyes, inedible but treasured for their dark mahogany brilliance, have been picked up and carried into the house and poured decoratively into a flat brass bowl—except one which will be carried for luck, polished and comforting all winter, in a boy's pocket.

In her brief idle pause, nature shows neither regret nor confusion. She has not lost sight of her aims in this all but imperceptible pause, and every one of her creatures understands that.

ON THE THEORY that example is better than precept, I went out yesterday to rake leaves.

This is a job that must be done slowly, in a reflective mood. Also one must first locate the rake. I found it, finally, under the pile of leaves raked up last weekend, so the visiting small cousins would have a place in which to practice standing on their heads.

Next one must lean on the rake handle, admiring the scenery, the magnitude of leaf-fall and one's own courage, the sunny autumn day, and life in general.

While my leaf mountain grew, I thought over some of the summer's events that occurred while those leaves were growing old. A tree's fiscal year begins with the separation of one crop of leaves from its branches, where already by that time the tight, pale-brown knobs of next year's leaves are formed, to swell and shrink all winter, according to the fluctuations of temperature and moisture.

Raking up long swaths I reflected that a tree works all year to produce this annual accomplishment, for me to scoop up and carry to the midden behind the barn, where leaves will grow soggy and disintegrate. For the tree, the leaves are like my daily chores, of meals, bed-makings, floor-sweeping, laundry, which take perpetual energy and leave no permanent record.

But the colored leaves now lying on the grass are not all. Inside the dark, rough trunk, the tree has added a new layer of live wood around its core. And something remains from a year of farm living, too: from things shared of good and bad; from utter laughter or the crackling of anger and clearing-up afterward; from meals eaten together and work shared; secrets told and disappointments comforted away; and from supporting loyally without first demanding an explanation. From these has grown a layer of strength that will remain as a permanent record, long after the tedious household chores are raked up and carried out to the midden to disintegrate.

CORNSTALKS from last summer's garden now lean toward the kitchen window, and the November wind goes through them in a shudder. Their thin tassels spread out beseeching fingers, and their long, bleached blades flutter like ragged clothing.

November's cornstalks are expressive of a human mood that genuinely needs the pause for Thanksgiving.

I like November because it is the month in which, four years apart, our boy and girl were born. I like it even more because it is the month in which Thanksgiving Day comes.

I like big, noisy, happy gatherings of people, and since a farm is the ideal place for them, the family gatherings usually happen here.

I like this shabby, worn, old house cleaned to polished hospitality, a fire leaping in the dining-room fireplace, the best blue china and all the inherited ironstone and the old cloudy-glass goblets set out in readiness. I like the kitchen brimming with the odors of many delicious Thanksgiving foods, and the guests' perfume and tobacco, and the smell of cold air when the door is opened. I like the sudden crisp chill as I run out to the fence with the last bowlful of vegetable peelings. (In a crowded kitchen there is no room for them to wait.) The women put on my aprons to protect their smart wool suits and dresses, the men go up to the barn with Dick and Joe to see the new calves. Somebody—usually Grace—

plays the piano and somebody else puts a record on the phonograph, and everybody talks at once. They exchange gifts, news, political opinions, flattery and unflattery. Everybody offers to help in the kitchen and a few stay; everybody gets in the way. Someone sets the dining table, counting and forgetting and counting again. Even going at top speed with the dinner, I hear and enjoy snatches of everybody's conversation, and carry on conversations of my own with everybody.

Everybody gets hungrier and hungrier, until finally dinner is ready to put on the table. Dick always urges me to put on everything at once "so there will be no need of this jumping up and running to the kitchen."

At the last possible moment the hot rolls are brought to the table, and there is a quietness around the table and then Stevenson's indelible words: "Lord, behold our family here assembled; we thank Thee for this place in which we dwell, for the love that unites us; for the peace accorded us this day. . . ."

Thanksgiving Thursday is not always the day on which these gatherings take place. Some families get together the Sunday before or after, because many come from long distances or have jobs they cannot leave on Thursday; the time is short, and traffic dense.

Frequently farmers spend Thursday of Thanksgiving picking corn, or they go hunting. Each can be thankful in his own way. This is freedom, and it, above all, is reason for Thanksgiving Day. For freedom is the most cherished of all of man's possessions, and the last to be surrendered.

Foolish bees

Traditionally, farming is a good way of life in America. A farmer hears himself congratulated often and in glowing words: "It's the independent life. You're your own boss on the farm." He accepts this with a smile and some reserve.

It is true. On the farm you are your own boss and can do exactly what you please, as long as that also pleases God and the weather; the government; the neighbors; your children and the neighbors' children and their 4-H clubs; the farm dog; the cattle and hogs; the insects; crop surpluses; the farm tractor, money inflation, and income tax blanks; the U.S. Army, Navy, and Air Force; atomic bombs, missile fallout; steel strikes; advanced math and science in high school. And a few other things.

Still, the sense of independence persists. The farmer does have a good life, probably as nearly independent as any life now existing in this civilization. He likes it.

When all the reasons are simmered down, like cider ready to be added to old-fashioned apple butter, farmers like farming because they like the kinship with the soil. It is ancient, persistent, and healing. Farmers like it enough to be willing to make all the sacrifices necessary to remain farmers.

A farmer doesn't know exactly how to say this. He is likely to be inarticulate, anyway. His way is shy, diffident. He says simply: "I'd ruther farm." And then he smiles.

LATE LAST SUMMER a swarm of bees left the old beech tree, which had been a bee tree for as long as anybody knows, and followed their injudicious young queen to a small hedge-apple bush no farther away than the width of a barnlot, and there began a new hive.

They hung five slabs of celled comb vertically from the thorny limbs of the sapling. The slabs varied in size from a turkey platter to a pickle dish. It was then already too late to fill the many hundreds of hexagonal wax cells with honey. There was no shade for the hive in the ferocity of late summer, so the lower end of the comb became warped out of shape. There was no protection from snow and biting winds to come.

The cold came early. By last week, when we discovered the hive, the colony was already doomed. A few dead bees still remained in the uncapped cells, but most of them, dead from cold or hunger, lay on the ground under the sapling. One lone bee was flying above the dark-streaked comb. The next day he also was gone.

Looked at edgewise, the pale-gray empty comb looks like a handful of transparent, thin snakeskins hung there. Looked at broadside, it appears a prodigious blunder made by a young, inexperienced queen.

IT WAS just after 8 o'clock on Saturday afternoon in Ellettsville (population around 800 normally, increasing to around 12,000 in September when the annual three-day fall festival occurs there). Some of the merchants had closed their stores; others were about to; one was walking out of sight down the street.

Dick suddenly remembered he had forgotten to buy a cap. "My old one is too dirty and greasy to wear to town or sales," he explained. "I think I'll ask Cort to open up again and sell me one."

He stepped into the intersection of streets and yelled at the departing merchant: "Cort, oh Cort!"

Nobody showed any surprise. Cort, formerly a farmer, came back, unlocked his store, turned on the light, and made neighborly conversation while Dick selected a cap and a pair of gloves.

Cort's store is a leisurely place that sells a great many things, and nobody is urged to buy anything. There was an assortment of men's and boys' clothing, dry goods, hardware, kitchen equipment. A stack of milk buckets, tin pans, and small tools were displayed carelessly in the window. There were the red and black plaid caps that are standard equipment for farm men and boys; the soft, warm, brown gloves; the stiff canvas gloves; blue denim overall jackets; assorted boots and overshoes.

Not all of the stores in Ellettsville are so casual. Before Cort retired from farming, his store was a barbershop, and it is characteristic of him that the room still has the unmistakable stamp of its earlier career. His selling counter has a cracked marble top, which is convenient for Cort to write on when figuring up a bill that is inconvenient to add in one's head. The towel bins and small drawers that formerly held shears and barber supplies now excellently hold bolts, assorted screws, nails, and small hardware. Presiding where the barber's chair had been, Cort smiled, talked, offered caps. His dark, shrewd eyes were friendly and unopinionated. Cort is accustomed to listen, think, and withhold his opinions.

At the back of the store, near the coal stove, was the "liar's bench," a long, white-painted wooden bench that came out of a country church. In summer Cort sets it on the sidewalk against the front wall of his store. In winter it is near the stove, and there is a regular gathering of friends who come and sit there, to joke and taunt each other and tell tales.

While Dick selected his cap, I looked at the wall behind Cort and could imagine the dim outline of the old-time shaving mugs that once occupied the high shelf, some daintily decorated with blue and pink flowers and a name in gold, some more conservative with orange bands and gold thistles and a name in gold. The big mirror, now freckled and dingy, still occupies the wall behind the marble-topped counter. But Dick, having selected a cap, did not look into the mirror. He put it on and looked into his wife's face. "Oh!" she exclaimed, "I think you look just beautiful!" His daughter nodded, and Dick smiled at Cort. "I'll take this one," he said.

❦ Last night's heavy frost loosened the golden leaves and this morning, when the school bus came, they were falling like confetti. Still it was noon before you could notice any diminishment of leaf on the young maple by the driveway. By that time the sharp-pointed leaves covered the ground under the tree, like ruffled gold feathers covering a bird's body.

Suddenly there sprung up a clamorous screeching, then a multitude of birds rose from the ground in the east field and flew into tall trees in the yard. They screeched as if applauding themselves, goading themselves to greater effort in some kind of violent game. They seemed even to have teams, yells, signals, and a goal. Occasionally some left one group and flew toward another across the yard. At intervals they all left the trees and flew swiftly, rising, dropping, circling, whirling, turning so suddenly that if they had been people they would have collided and the ground would have been strewn with dead and dying. Being birds, none collided, none fell.

They also took time to groom themselves, pecking at the skin of their dark bodies. They were starlings, identifiable by the long beak, short rumpled tail, small polka dots on underbody and wing. Their graceful, dark heads, on long throats, were somewhat like pigeons' heads. They are undesirable birds, but their gusty enjoyment of life is a delight to observe.

ALL NIGHT the moonlight was as white as frost, but by morning real frost was spread over everything. The frosty grass twinkled; frosted small leaves, clinging to the spiraea bush, were stiff and glittery, like flower petals dipped in sugar.

The late-hatched red and yellow Bantam rooster, having

crowed early while the frost was still merely moonlight, now came out to make his weather observations. He stood on one spiky yellow leg, holding the other up against the cheerful yellow of his feathers, while he contemplated the situation in deep surprise. This was his first experience with the new kind of moonlight that settles icily on things in the unguarded late hours of night, and he was going to take his time about handing down an opinion on it.

❦ TODAY was the day I had set aside to do something about the green tomatoes.

I knew on that frosty, sentimental, garden-farewell evening two weeks ago that I was saving too many green tomatoes. There is a limit to the amount of green-tomato pickle a family can eat; even to the amount of green tomatoes that can be stored in the basement and kept, ripening, until Christmas. Besides, this house has no basement.

But, as I went through the garden that cold evening, I had felt an appeal from the old vines that had been so upright in their youth, so generous in their middle age. That evening, limp of leaf and stalk and sprawled flat on the ground every which way from the center, they had seemed to beg me to accept their final gifts. I could feel the plea from every hard, shining, frost-threatened green tomato: "Save me, save me. Here am I; take me."

So I gathered nearly six bushels and carried them into the woodshed, covered them heavily with gunny sacks, and hoped the day would come when I could use them.

It's not so nifty to be so thrifty

Last week I noticed juice had already begun to ooze out of the baskets from the split, bruised, and bug-bitten tomatoes. They had a smell of urgency.

I got out my canning books and looked up recipes. I found one recipe for uncooked green-tomato pickle that uses a peck of green tomatoes, and one that uses a gallon—but it also takes onions, horseradish, hot peppercorns, celery seed, mustard seed, and other things. There was a recipe called "tomato chunks" that calls for slightly ripe tomatoes, and one for Southern crystal pickles that can use a gallon and a half of green tomatoes—but it also requires slaked lime, stick cinnamon and ginger, and more things. There was one recipe for onion and tomato pickle that would use a gallon; one for tomato cocktail pickle that would use another gallon. There was a recipe for green tomato mincemeat that takes eight pounds of green tomatoes, chopped or ground. You add spices and can it, and add meat later when you bake the pies. There were two chowchow recipes. I abandoned the one that uses only two quarts of green tomatoes in favor of the one calling for half a peck.

Finally it was clear that six bushels of green tomatoes had been too much. Some families like fried green tomatoes; this one does not. "I never was so disappointed in my whole life," wailed Carol the evening I offered them. "They looked like the most wonderful slices of delicious meat!"

I began sorting the tomatoes. While sorting, I made up a recipe for green-tomato casserole that I hoped I would be able to use maybe once a week all winter. In my imagination I sliced the tomatoes, salted and stacked them to drain as you would eggplant. Even in my imagination I had to hide the crock so nobody would discover and throw the slices away. In real life I wiped my hands on a burlap sack and started sorting the second bushel.

When the imaginary slices were floating in pale green brine presided over by vinegar gnats, I drained them, dipped them in cracker crumbs, and laid them around a roast already done but not quite browned. In my imagination the light was too dim to see whether the roast was beef or pork. Starting in reality to sort the fourth bushel, I added secret seasonings to the imaginary slices. I used powdered ginger, slaked lime, onion salt, celery seed, dill sauce, and horseradish. At last in my imagination I served the whole deliciously browned achievement on the big old pink and green ironstone platter (the fifth in-reality bushel was of half-ripe tomatoes). I gave it an exotic foreign name that nobody could translate.

Then, at the top of my imagination, where realism is a crisp brown crust, I heard somebody ask: "What is that stuff? Could I have a peanut-butter sandwich, please?"

I removed the burlap from the sixth bushel, and that bushel I took out and resolutely threw over the fence to the hogs.

❦ GRAY RAIN was falling from a gray and linty sky upon gray leaves lying on the ground. On the front porch the old summer rocker kept getting darker from rain. Its torn cane seat bristled as if it, like the cattle, were growing long winter hair.

In the field across the road, beyond the reedy gray curtain of dead weeds, the neighbor's cattle began coming in from pasture, walking slowly, single file. They could have shortened the walk by going straight uphill, but they had plenty of time, and good judgment beside, so they were winding around on an easy level, making a gradual ascent.

The cattle were not being driven in by bad weather; this was their customary time to come in and this was their customary gait, rhythmic and placid. A paired lifting up, putting forward, and setting down of their feet; a rocking forward of their bodies, causing their blocky heads to swing loosely, chins bobbing, as if the heads were mechanically suspended, well-balanced, and effortlessly held at that level.

Fascinated by their appearance of content and leisure, I watched them several minutes and, when I returned to my own rainy-day routine, discovered I had absorbed some of their relaxed mood. This is a profit you can take, without investment, from your neighbor's cattle.

SOME WINTERS come suddenly, like a hawk pouncing on an unsuspecting fat hen, but weeks ago this one sang out a long get-ready signal: "A bushel of wheat, a bushel of rye, all not hid holler 'I'."

The farm began to get ready for winter. Walnuts fell. The beech tree threw down the last of its three-cornered brown nuts from their pronged hulls. Chipmunks insulated their underground halls. Farm animals put on their long coats. Birds went south. Trees shed their leaves, except for a studious few that kept their somber colors like a lamp burning narrowly when someone bends close to a book late at night. Days grew shorter; nights came earlier. Even at midday the sun came through the dark branches in the bluish tones of winter.

Winter approached, singing in its cold-wind voice: "A bushel of wheat, a bushel of clover, all not hid can't hide over!"

Suddenly yesterday the wind sang out its final warning: "Here I come with both eyes open."

The day was cold and windy, the sky heavy and gray. We had to keep the kitchen light burning all day. When I hung wet clothes on the line, I wore scarf, gloves, and winter coat. The sheets raged, arguing with the wind for a while, then subsided in frozen dignity. From the maple in the front yard a broken rope swing dangled emptily.

As I hung up clothes, I could see the cornfield up on the hill, from which all the corn had been gathered now. Above it, high overhead, passed a late band of wild geese. I felt a deep sympathy for these eleventh-hour travelers.

Everything knew the snow was coming. I knew it because the radio had told me; wild creatures knew by instinct.

The sky darkened; the wind hurried; suddenly the snow came. Against the dark background of barn and woodshed the long, white lines were driven fast, at a sharp angle, upon the yet unchilled earth. There the first wet, soft flakes melted quickly, but as the ground cooled, they survived in little patches of white. Snow fell faster, more densely; the flakes became larger. The wind died down completely.

As I turned toward the house, carrying the empty clothes basket, the snow was suddenly all around me, like a great white bowl turned upside down, shutting out the rest of the world.

Winter had come, and the farm accepted it gladly. "One, two, three for me. Home free!"

DECEMBER

[DECEMBER]

❀ ❀ ❀ ❀ FOR a farmer who has the luxury of brush to burn, a windless bright day in early December is an added blessing. In this farming era, when soil conservation is akin to godliness, no farmer cuts down trees merely for the pleasure of having brush to burn, however much he enjoys it, but there are times when brush is on a field and ought not be, for the sake of tillage or soil cherishment.

On such a morning, having run quickly through his barn chores taking care not to get involved in any big ambitious projects like repairing the hinges on the corncrib roof, a farmer fills a small prune-juice bottle with kerosene, gets a few matches from the kitchen, and picks up the ax.

The ax, because if there is a limb big enough for fence post or bean pole, it must be trimmed and laid aside for that use.

The joy of being alive on such a morning leaps quickly into flame in a farmer's spirit, but the brush fire takes hold slowly, small and coaxed. By the time it has yielded to the temptation of kerosene, the man's spirit is warming itself in the glow of familiar surroundings—sun shining, sparrows twittering, crows cawing, dogs yapping in the distance, cows bawling in comfortable curiosity, having smelled the smoke from the the first burning twig.

Frost sparkles on grass and brush pile; sun sparkles on planes passing high above the field. The planes are less of an interruption than the occasional rabbit or quail that shows itself beyond the brush pile.

By the time the fire is a great red mouth, biting and chewing up brush with a cracking sound, the farmer has long since shed his denim jacket and his thoughts have gone on a wide, evaluating tour, which is the bliss of brush-burning.

From his field he can see the neighbors' farms and some of their barns and houses. It is a pleasant closeness, without trespass. The farmer reviews the local, unprinted news and his winter work plans. He thinks about where his son will go to college next year, and about his wife's supper menus of late. He reflects on his neighbor's political summaries; the church's possible enlarging, the increase in county taxes; the world news heard that morning on the radio, of which he will read more when he goes to the house at noon and the mail has come, with the morning newspaper.

Brush-burning

For all his seeming isolation out there burning brush, a farmer is in the midst of the modern world. It is a place where anger crackles like the sticks he is thrusting into the fire, where fear and compassion take flame, and hope rises like smoke. Hope of peace, hope of survival, hope that the country's leaders will speak up fearlessly, savagely if need be.

The farmer's anger flames at the morning's discussion of people going underground, like a race of moles, for survival. And the evacuation of whole cities—who would go while the children were left behind, in school? There wouldn't be room for all the people, the farmer had insisted angrily; the roads would be clogged with confused and vulnerable refugees. Why not stand fast while there is still time for reason and sanity, and refuse to yield up the things our forefathers deemed precious beyond life itself?

By the time the farmer remembers to look at his watch, its hands say it is noon. His clothes and hands are fragrantly smoke-saturated. He lets the fire subside to a safe red core that can either die out entirely or be rebuilt easily after dinner.

If he doesn't get back that winter, or for several winters, the brush pile will make a brushy, safe home for rabbits and a few birds, and some other small animals that need homes. It is really better for such use if the brush pile is near a hedgerow; but a brush pile out in the field makes a good oasis for small animals that have to cross from home to hunting ground and do not like to risk their small, important lives traversing too great an expanse of unprotected open space.

❁ IF THERE's anything more irritating than the sound of a person eating an apple, it's the sound of him trying to eat it quietly.

❁ WE CAUGHT a possum in the henhouse last night. He came in through the door, which nobody had remembered to shut before bedtime. When Dick got to the henhouse, having heard the hens' wild clamor, the possum was crouched as far as possible back in the corner, trying to get out of the lantern's light.

When he returned to the kitchen, rain was dripping from Dick's cap and hands and he was dangling the possum by its bare, ratlike tail. It was sleek, fat, and unperturbed. Without seeming to struggle, the possum kept twisting his head sidewise and arching his body upward.

"He's trying to climb up his tail and bite my hand," explained Dick, shaking the captive down again. We looked at the long, sparse, black and white hair; the narrow, piglike snout; the toothy grin and little unfrightened eyes.

"I reckon I may as well let him go," suggested Dick.

"Will he come back?" I had fed and cared for those chickens since early last spring, when I had to get up at night and go down to the brooder house to make sure the stove had not gone out.

"Oh, yes, now that he's got onto it, he'll probably come back every night—unless somebody remembers to shut the henhouse door."

It is hard to sentence a healthy creature to immediate death. Besides, the possum had killed for food, not for sport or cruelty, and he hadn't even got to eat his supper. His only sin, actually, was that he blundered. On the other hand, a farmer's duty is to defend his stock.

The world in which one deals with animals is a kinder world, in some respects, than the world in which man deals with men. With animals, man can be inconsistent. He can compromise, dispensing mercy through illogic and pretense, and thus avoid a too-final, difficult decision. With men, he can be merciful, but he must never compromise.

We put the possum under a washtub and went to bed.

In the morning, since he was going to town anyway, Dick took the captive along and tried to give him away. Nobody wanted a possum for any reason at all. It would have been unneighborly to release him on a neighbor's farm, so Dick brought him back and turned him loose in our hilltop field across from the church, paroling him to a cow that was grazing there. Whereupon the family, in council assembled, passed a law that hereafter the henhouse door must be shut and fastened by the first person passing there at dusk, and double-checked before bedtime.

❁ ON INDIANA FARMS you see piles of yellow corn dumped into makeshift cribs surrounded by snow fence.

This is the farm surplus and will be used before the corn stored in the regular, roofed corncribs.

There lies summer; the long hot hours of July and August gathered and held in the tapering, golden ears. Fed to livestock, it will again become heat and nourishment, uniting the seasons in an endless cycle. Corn is the symbol of farming—the backbone and richness of it.

What a tragic inconsistency that this golden abundance should also be the cob of a continual farm problem, especially when there are countries in which this surplus could mean the difference between starvation and life.

There must be a way to correct this unbalance. There is; but it must be delicately done in order not to destroy something more vital than a fed stomach. A greater law than digestion is involved, a law that is constantly repeated in the natural world.

It is not a question of how much you can give, but of how much you can accept and still remain free.

❁ ALL THROUGH the cold, starry night, there was the sound of wind running restlessly, like a colt exploring a strange pasture. It brushed the empty branches of trees and set them to dancing. It pushed the torn-cotton clouds back and looked inquiringly into the bright, serene face of the moon.

It was as if, in December, autumn had returned for a postscript.

As I came down the hill, late at night, I saw five stars

fall. They were orange-colored like fragments of burning leaf. The first one fell alone from the north; then two fell together farther out, as if racing each other to the ground. The fourth one fell alone, slowly; and then the last one fell. All were like the tossed-up, thin embers blown from an autumn leaf fire, which one watches to make sure they die out safely.

❁ YOU DON'T see many guineas on the farms anymore, but for that matter you don't need many to make all the noise you can stand. Guineas are instinctive vetoers; they protest every change in their surroundings.

The seven spring-hatched guineas here were accustomed to Dick in his blue work shirt and overalls, and when he appeared this fall in a jacket, they were bitter in protest. When I walk across the yard now in a pink coat, they scream from the barnlot. Usually guineas will not roost in the same house with chickens, but Joe's guineas willingly share the henhouse with his Bantams and ducks and Albert, the goose.

Now I know what the guineas say, crying excitedly in their foreign-sounding language, when I open the banty-house door. They immediately explode out of it, take a few running steps, and become air-borne over the raspberry patch, revealing the whiteness under their gray and white, polka-dotted feathers. "Lookit the wreck, lookit the wreck, lookit the wreck!" they scream, giving me the conviction that they mean me.

Still, I like them. From a distance they appear to be

wearing fitted bodices and full, black and white skirts. From close-up they look foreign.

Joe takes pleasure in exhibiting his various poultry to people visiting. He brings down the gold-flecked Bantam hen, serene and quiet as a sunny morning; or a gorgeous wild mallard duck, Rembrandt-colored in purple, dark blue, green, and white; he brings the one big white hen that lays the beautiful coffee-brown egg; he brings the guineas. All these recognize him as their friend, permitting themselves to be touched, admired, and researchfully examined under his sponsorship. This is a remarkable thing for guineas to do; they are wild, they have strong muscles and must be held firmly, even though in Joe's hands they sit as still and immobile as the old sawed-stone corner posts in the fences on this farm.

Last week my sister Nina was here, and Joe brought a guinea into the kitchen to show her. He held the gray and white taffeta-feathered body on his palm, but his hands gripped its strong legs unrelaxedly. The guinea's round, bright eyes never moved nor closed, although the bird was fully alert to all that went on in its unaccustomed surroundings. He took hold of its long, thin neck and drew it out to the full length from the plump, wing-folded body. In profile, a guinea's back is slightly humped, like the back of a flea or a grasshopper. The small, bare, turkeylike head has wattles and ears of paraffin-white and bright red. The wrinkles in the long neck were of the bright blue of copper rust. A few straight strands of black feather stood up raggedly behind the red and white cockade, slightly suggesting a war-painted Indian. The gray and white feathers, each one individually polka-dotted, merged, in the mass, into gray and white designs of extreme beauty.

Nina, charmed, gently lifted and unfolded the taffeta-feathered wing, from which one expects a dry taffeta whisper when the guinea walks.

The most fascinating discovery about the guinea was its control of neck muscle. No matter how the neck was bent, the head remained fixed at an unchanging level, like something floating in air. On the end of the long, wrinkled, blue neck, the guinea's head extended eight inches above Joe's palm, and no matter how he wafted the guinea's body, whether up or down, to right or left, or even in a small circle, the head retained its same, exact latitude and longitude, as if it were in no way concerned with what its body did. It looked as if you could have wrung its neck, without even touching it, simply by turning the body around and around steadily.

"Why, that would get attention even on television!" exclaimed Dick.

❀ THIS MORNING the pastures have a brown and faded look, like an apron that has had too many careless washings, but enough sturdy green patches remain to keep the cattle grazing. The naked trees are dark against a gray, noncommittal sky, and I have the conviction that somewhere snow is being ground up for early delivery to the farm.

The drabness gives the farmscape a quietness that is not at all unpleasant. After the glory and eloquence of autumn color, one really welcomes a moment when there is nothing calling for applause. Now the sudden flight of a small procession of English sparrows past the window is drama enough.

❁ AT DAWN the air currents begin to stir and the general temperature becomes a few degrees colder. I used to think this might be my imagination, but when I mentioned it to my neighbor Iris Stranger, who used to be a geology teacher, she said: "That is true, according to the geography book."

This gives me further confidence in something else I have observed: just at dawn the earth gives off a changed fragrance. The difference is subtle but discernible. It is fresh and pleasant, and created, probably, by the mingled fragrances awakened by the freshly stirred-up air currents that also create the coolness.

❁ IN DECEMBER the new calendars come. The last pages of the old ones are not used up yet; the new one seems to shorten the year prematurely.

Unrolling the new calendar from its wrapper, I look around the kitchen for a vacant nail on which to hang it. There isn't any.

The new calendar crackles an energetic greeting, like a new sales manager sweeping clean: "Well, what did you get done last year, anything at all? Get the barn painted? The roses moved? My, how fast you're getting gray!"

The old calendar is breathing its last, and I know I shall have to make friends with the new one. There are, actually, several new ones, but from the moment the first is hung on the wall, the weeding-out process begins.

Survival of the fittest applies as truly to calendars as to farmers or high school basketball coaches.

Calendars come in a variety of sizes, and by diverse ways. Some come by mail, in odd-sized envelopes or rolled up in a tight wrapper. Some are handed out by salesmen at the door or in stores. Some come importantly in cardboard tubes. Some come naked as from the press. They have one common purpose, to perpetuate in kindness the memory of the donor in time of money expenditure.

There is the calendar with the big picture and the little page. It is the first to come off the wall. There is the one with valuable information on the back of the page, but it is never consulted because it has to be taken down from the nail every time you turn to the new month's page. About July, time comes to a standstill on this calendar. There are calendars with splendidly colored pictures that come rolled up and do not flatten out until May. There is the glued-page type; when you tear off February the pages give way up to November. Nobody can hope to keep up with such a swift passage of time, so that calendar goes.

The one that remains longest on the wall is the one with wide-open spaces around the dates so that memoranda can be written on it. Its white spaces fill up with notes of dental appointments and Fair Board meetings, who called about baling; with telephone numbers, grocery lists, addresses, recipes taken down over the phone, and important farm news such as "barn swallows returned" or "paid on tractor."

It already has its own reminders, such as "Robert E. Lee born," "First airplane flight," and the moon's phases, times of rising and setting, the probable weather, and all holidays. When this calendar is used up, it is too valuable to throw away. It belongs with the farm records and receipts and will be called on to help fill out income tax blanks. It's a diary, a volume of family history.

❁ IN THE NIGHT "it weathered," as farmers here say, and by morning the back-porch window had become a splendid thing, a winter etching done in frost against glass. Long lines of bristly, Christmas-tree tinsel were crisscrossed into delicate cobwebs, designs of stars, scenes of houses along a narrow village street, a little church with pointed steeple puncturing a starry sky, deep forest, still fields, mountains.

Not an inch had been wasted. The little spaces between the silver pictures were filled in with tiny bracelet charms in the shapes of animals, fruits, leaves, many-petaled flowers, old coins.

The early morning sun poured a lapful of diamonds over the silver etching and the whole window glittered in almost unbearable splendor.

So it seemed to me, from inside the room. Outside, a little gray junco came in the bitter air and chirped distressedly against the splendid window, until I went out and tied a piece of suet to the nearest peach limb. Then a whole flock of these little gray birds came and perched there, like gray peaches on a winter tree.

The wisdom of birds impresses me acutely today; that they arrange their hatchings so as to have no little ones to feed at this time of winter.

❁ WHEN DICK brought in the milk this morning, he said: "Twelve above. There's not enough wind to disturb the smoke; it's just going straight up from the chimney. A fine winter morning."

He went back to the barn. I strained and put away the milk and went ahead writing two hours of letters that had to get out in the morning mail.

As I came up from putting them into the mailbox, his voice floated down from a high place near the barn: "Say, can you come up here and get me out of a trap?"

"A trap" could mean that a neighbor was there, trying to coax him to go to a sale that day, or that someone was on the verge of selling him something we can't afford. He didn't sound worried, nor in pain, and as soon as I got up to the barn, I discovered "the trap."

He was, literally, trapped in the silo. To put down silage, he pushes the wheelbarrow under the narrow wooden chute that is attached to the side of the concrete-block silo. The row of silo doors (small wooden squares that are lifted out from outside) is in the silo wall, just inside the chute. He climbs up the chute to the first opened door, gets into the silo, and with a pitchfork throws silage down the chute; the silage falls into the wheelbarrow. That morning, when the wheelbarrow was full, he had just kept on pitching down silage and the chute filled up, too.

Fortunately, the silo was still nearly full—just a comfortable height. While he stood there waiting for me to come out to the mailbox, he was able to rest his elbows on the rim, with his head and shoulders visible above the top of the silo. He could see all over the countryside. Patience was all he had needed.

We took time to laugh. "You look simply wonderful up there, Napoleon," I said. I took the pitchfork and began to fork away silage so I could move the wheelbarrow. "Why did you throw down so much?"

"Why, it was just so easy and pleasant. I got to thinking and forgot to stop."

I stopped forking. "Who were you thinking of?"

"I was thinking of Carr Stanger. I can see his barn from here."

"Oh, all right then." But it was an opportunity that doesn't happen often and must not be overlooked. "I'm tired now. Anyway, this isn't woman's work. I'll come back this afternoon and dig some more. I'll get you out in time for supper."

He went on, unperturbed: "I've been watching the countryside. Ralph's dog came down and played around for about twenty minutes, but there hasn't been a car this way all morning."

The silage piled up in the feedway and fell into my boots, but in a few minutes the chute was clear and the farmer came down like Santa Claus from the chimney. He immediately picked up a long, pickled straw and began chewing it.

"I've been in worse places," he said.

"Anyway," I agreed, "if I hadn't had to go to the mailbox, you could have eaten your way out in a week or so."

❁ THERE WAS SNOW on the ground and the evening was darkening when I heard the distressed chirping of a young chicken at the barn. The little speckled hen had chosen that evening to teach her sixteen fall-hatched chicks to roost high instead of sleeping in the apple crate that had been their nightly lodging place up to then.

She had chosen the long board that sticks out above the bull's stall. There was room for all of them on the board. By clucking, whirring, and spreading out her wings to their maximum tender wideness she had coaxed all but two chicks up beside her. Still on the ground were the yellow one and a small white one no larger than a half-grown quail. The yellow one tried several times and finally achieved the board, but the white one was disconsolate. He was pulled by a longing for two different, familiar comforts, his mother's wings and the apple crate.

Finally the apple crate won. He turned and went out

doggedly, a little white blur in the dark, toward the crate. There I gathered him up and handed him to Dick, who tossed him up lightly to the hen. He lit on the board; there was a moment of uncertainty while the whole group wavered and regained their balance. None fell. The little white chick nestled under the mother's wings and stopped chirping. After a few more nights he will be reconciled to the new place. Then the hen will leave them all to look after themselves.

❁ IN A SPIRIT of good-natured research, which Dick calls "rootin' around," the red Duroc hogs accomplished what no one else in the family has been able to. They persuaded Dick to give up his old brown wool coat. They did it simply. He left it on the tractor seat when he came to dinner, and they ate it. The biggest piece left was no bigger than a blouse pocket.

Dick will be just as warm, and much better dressed now, in his second-best overcoat, but the loss of the brown one was a sorrow to him. It had raglan sleeves and no belt, and had belonged originally to Teddy Schneider, famous accompanist of famous tenor John McCormack, and had come down, through a series of dear friends, including his bass-voiced younger brother, to Dick.

It was one of those large, loose garments that fit almost anybody. Except our neighbor John Dunning, who tried it on one day. The lining stretched, protested, split apart, but that did not in any way detract from its charm and usefulness to Dick. It came down a few inches below his knees, and where it was torn it came farther. He tied a piece of baling twine around his waist because "Uncle Howard always said a belt is equal to another coat, for warmth."

The hogs had no other motive than their natural passion for research.

Last summer when Ira Stanger was here, pointing up the silo walls, one of the hogs started to carry off his whitewash brush. Dick gave chase, rescued the brush, and on the way back met another hog carrying off Ira's trowel. Last week, when Clyde Naylor stopped here with a broken hay rope for Dick to splice, the men stopped to talk and the hogs immediately started carrying off the pieces of rope.

The day after the coat affair, they resumed their research somewhere out on the farm, and as if in compensation, brought Dick in a little gift, which they left in a pile of bright red corncobs in the barn. It is a little white milk-glass lid, from some long ago mustard jar. About four inches in diameter, it represents a nest of hatching eggs. The knob is formed of one chick, peeping out of a broken shell. There are nine eggs, some pipped, some with chicks hatching out of them; two are going to prove infertile, obviously. Washing off what seemed to be persistent dirt, I discovered it was gilt paint. How the lid got buried, or where it was when the hogs researched it out, or why they didn't break it carrying it to the barn, or whether they will ever unearth the dish that goes with it, invites the imagination.

"That's the beatenest bunch of hogs I ever saw, to pack things," declared Dick.

❁ WHEN I WENT through the snow down to the mailbox this morning, there was a line of footprints reaching from the front porch down to the road. The space between each two steps was as long as seven ordinary-sized steps. Obviously Joe, having lingered too long over the story of the Wright brothers' first plane, had to make a wild dash for the school bus.

By the stone post near the edge of the road, having gone in ample time and making her small footprints, Carol had observed a line of animal footprints across the yard and had labeled them, accurately, "rabbit."

As I came back to the house, I took time to label the giant prints, "boy."

❁ THE FIRST IMPACT of the Christmas music is almost painful in the poignant stirring up of memory and sentiment. The carols never lose their evergreen ability to invoke Christmas.

But I wish the stores would not start playing Christmas music before Thanksgiving. Carols belong inviolably to their own season.

❁ LAST SUNDAY, unexpectedly, it became my chore to take a petition along a five-mile length of the Maple Grove

Road and ask for signatures. Dick, who much preferred taking a nap, insisted on going with me because "I can save you time by staying in the car and honking the horn when you take too long."

In some houses dinner was over and the tables cleared. In others it was still a ready-mix just poured out of the box. At one house a big family reunion was in process and everybody was cooking something. Some houses were askew with Sunday's relaxedness; some were as neat as a new book. Some were decorated for Christmas. In the Brown's 128-year-old stone house Christmas music flowed softly through a room beauteous with gold leaves and colored fruits and painted candles. At a new stone house a small, pale girl stood by a window, solemnly arranging and rearranging the Nativity scene there.

At one house a surprising thing happened.

Once a dear neighbor lived there, but later new people moved in, who seemed suspicious and unfriendly. For more than a year now I have passed that house with anger and resentment because the man who lives there had done a harsh injustice to Dick and Joe. I almost decided not to stop, but it was a civic duty, so with a chip ready on my shoulder, I did stop. The man and his wife both signed the petition willingly. Then the man said, like a person throwing down an armload of wood, "I want to apologize to your husband and son for what I said that time. . . ." He explained why he had been angry and unreasonable. I admitted we had been unhappy over the accusation and would be glad to have the unfriendliness done with. He invited us to come and visit them. He meant it sincerely, and I meant it sincerely when I said we would.

It sounds like the typical, trite-plot Christmas story, but how heart-warming it is, in real life!

❀ WE SET UP the Christmas tree yesterday—a five-year-old Virginia pine—and Carol decorated it as she wished.

When the children were little, we used to go on the tractor to Blaine's woods and cut a small cedar for our Christmas tree. When we brought it into the house, cold, sharp-needled, and usually with a bird's empty nest in it (because the children hunted until they found one that did have a nest), it gave the room a delightful cedary fragrance.

Now that the children are older their holiday activities extend in a wider range (such as the 4-H caroling party when the Maple Leaf club and sponsors go in Warren's truck to carol for the elderly and shut-in of the community as preliminary to a party). We buy a tree, although Dick has never really got reconciled to paying money for one.

We were driving along old Highway 37 yesterday afternoon when we saw the sign advertising Christmas trees, and a long row of them freshly cut and piled greenly on a grassy bank at the Lilly farm. They were seventy-five cents apiece, and Dick suggested it might be the place to get one.

We began the ordeal of making a choice. It is hard for Carol because she is hard to please—she would like to have a tree as tall as a giant redwood, she is critical of every branch and needle. For me the choice is hard for a different reason. It's like choosing one dog from a whole kennelful. Every one looks at me with pleading: "Love me, take me home." If I look at more than three, I am lost. My way is to take the first one I touch. I picked up a tall one at the first of the line; the gummy juice from its trunk came out on my bare hand. "I like this one," I said.

"I don't," said Carol. "It's crooked. It's almost yellow."

But we took it anyway. I gave the tree seller seventy-five cents and he thanked me in a pleasant, Scotch-accented voice, adding: "And I hope you and yours have a good Christmas day."

Carol got into the back seat of the car, intending to sulk all the way home, but unfortunately for that plan the tree was so tall it had to have the whole back seat and both back windows had to be rolled down besides to make room for it, unbent. She had to sit in the front seat with us and soon forgot to sulk. When the pine was set up in the living room and decorated with colored lights, shining metal icicles, and the fragile Christmas-tree balls and birds that have shared many Christmases with us, it looked beautiful.

After it had stood in the warmed room a short time, it began to give off a delicate, flowerlike perfume. When in the evening the tree lights were turned on, warm against its long, paired needles, the fragrance deepened. This perfume, characteristic of Virginia scrub pine, was a delightful surprise from the Christmas tree.

This is the Christmas tree I shall remember longest. Its tiny cones are rough and turn backward against the sticky bark. It is tall and outreaching, dense enough to hold decorations and small gifts in its branches. The branches come out of the main trunk at intervals, in an arrangement of five branches not quite in a circle, and the smaller twigs come out of the branches in the same fiveness, suggestive of the five points of a Christmas star. The Virginia scrub pine seems born to be a Christmas tree.

But there is another reason why my heart warms toward this fragrant, not ungainly, green tree. It will grow on abandoned or abused land, where other trees will not grow. At first only such plants as cat brier and wild honeysuckle, which like a sterile and acid soil, will grow where Virginia scrub pine is. By the time it is about five years old, it has made its own seedlings, holding the precarious soil in place and adding fertility until finally, after twenty-five or more years, the spring-lovely redbud and dogwood appear in the pine thicket. When many generations of Virginia pines have come and gone, the soil is fertile enough to support other, better forest trees. In seventy-five years almost any forest tree or plant can subsist on what the Virginia scrub pine has made possible.

❁ IT WAS my intention to make a molehill out of the mountain of unaddressed Christmas cards on my desk that evening, but the music supervisors changed all that.

"We wasted two hours and fifteen minutes practicing where to sit on the bleachers," Carol said, "so tonight Mrs. Garton wants us to come to the high school auditorium and practice the songs." Her face shone like a Christmas candle at the prospect.

Thank you, Mrs. Garton, for a lovely evening

This year the township schools will have a joint program, planned by Mrs. Joe Garton with assistance from her husband, who supervises music in an Indianapolis school. He will teach the township children how to stand up simultaneously and how to sit down without clatter.

We had car trouble and so arrived late. The bigger children on the bleachers, and the little ones in the long double row on the floor in front of the stage, were already singing when we arrived. Carol got unobtrusively to her place on the bleachers, and I sat down among the parental faithful in the audience.

The rehearsal was delightful, with some features that will not be a part of Thursday evening's performance. On Thursday the two evergreen trees on the stage will no longer be bare of Christmas ornament; neither will one topple over suddenly upon the heads of the surprised little second graders. There will be no repeated murmur of "Let's try that over, shall we, and this time remember. . . ." The little blond first-grader, fascinated by Mr. Garton's directing, will not imitate him. The three Kings of "Orientarre" will not be wearing plaid shirts and blue jeans. Mrs. Garton will not flit from one piano to another, asking which one do the children think brings out their voices best. The big girls will not wince in their own high note of *Oh, Holy Night*. The children will not be asked which color spotlight they prefer. (With one voice they cried: "Red!") Mr. Garton will not snap his fingers and tell them: "I want to see every eye on me." Mrs. Garton will not call the roll at the last to see which children, having been faithful unto practice to the end, are entitled to be in the program.

Thursday's performance will be poised and polished, shining like a great star, no doubt; but when the smallest children stand up and sing, "Wind in the olive trees, long, long ago," it cannot possibly be more poignant than it was at the rehearsal, with the half-light on their uplifted small faces, tired but earnest and pleased, their voices clear and grave and young as Christmas itself, singing an old song. It was a Child who gave us Christmas; it is children who keep it fresh. To get the real good of Christmas, you have to see it through the eyes of a child.

❁ "IF I DON'T survive this," said Dick when I had refused to pick out and wrap his Christmas gift to me, "I want my epitaph outlined in Scotchlite tape, saying: 'Merry Christmas, now you see him and now you don't.' "

❁ IN THE FARMHOUSE, electricity, running water, and automatic heat are acts of God, no less; yet there are times when by reason of these very blessings one could easily miss some of the finest hours of country living, and never know it.

The cold winter night, for example, when snow sparkles in the moonlight, and shadows under the telephone lines are strong enough to pull a sled with, and the diesel train whistles in a far-off, lonely wail.

Even more so, those last three minutes of very early morning, before darkness leaves the earth. The farm is covered with an inscrutability made up of many layers of black chiffon. All the familiar landmarks—the fence posts, bushes, trees, stones, buildings, the cistern pump—are enclosed in a dark anonymity. The earth is still dark when the first pale line of color parts the earth from the sky along the west side. The color deepens rosily, rising above the dark blue line that will presently become a wooded hillside on the far horizon.

I like to turn back quickly, before the veil is lifted and the identity of morning is confirmed in the lumpy, rough prints left by yesterday's feet across the yard.

Besides, it is too cold to stay any longer.

As I turn back toward the kitchen, shivering, I look always for the one large silver star that hangs in the blue sky, seeming no higher than arm's length above the roof. A ribbon of wood smoke comes out of the chimney, rising whitely into the sky. Its fragrance blends with the smell of coffee and frying bacon to evoke an earthy sense of pleasure.

My whole day goes the better for this brief, stimulating sip of cold, dark morning. I have acquired a taste for it, so that to get out of a warm bed in time for those last three minutes of darkness does not require undue self-discipline. It is a pleasure that can only be had at its own deadline. Half a minute too late is a whole day too late.

❁ AS CHRISTMAS draws nearer the days become spinningly more crowded with the activities of preparation: shopping, wrapping gifts, writing cards, decorating house and church and school, delivering Christmas baskets, the gaiety of parties, cookie-bakings, church and school programs, homecomings of big families.

Finally the days become so full and commercialized that there is no room in the inn for any more thinking or straightening out of one's personal beliefs.

Then comes Christmas Eve.

Everything has been done that there will be time to do. Now we have to leave the world to Christmas.

When finally the last carols have floated away into the starry darkness, the last good night has been said, the last wrapped package laid under a shining tree, there comes a small, quiet enclosure of time for every person to be alone with Christmas.

Why did we hurry so? Why did we worry about anything? What need we fear in the year that is about to begin?

Now in the stable it is quiet and blest. There is room for everyone now. Jesus is born, love is born, and the stable itself has become an inn.

❁ THE DAY AFTER Christmas we were having breakfast at the usual earliness and saw a rare spectacle—a colored end, without any rainbow attached, hung up in the sky by itself above the clover field on top of the hill. No rain was falling or even threatening. The colors were distinct and ran in all the prismatic shades from orange to lavender, and the segment grew steadily more bright for at least ten minutes. It reached steadily upward, but never attained the curve necessary to make it a rainbow.

That was because it was a sundog, not a rainbow, and it means that extremely cold weather is about to come.

❁ ON SUNDAY during the twelve days of Christmas Poor Richard and his wife were bidden by our neighbors, my Lord and Lady Telfer of Fair Dodhead Farm, to join them at the Boar's Head dinner, on the I. U. campus. It is a merrie old English tradition, held in the Union building, with madrigals and strolling carolers; a punch bowl, a boar's head, and flaming plum pudding being borne in to the sound of trumpets.

Accordingly, we did array ourselves in our best, I in my gold-colored woolen and earrings, he in his gray suit, and did hie ourselves betimes in our carriage to meet mine host. Lady Telfer was becomingly clad in rich, dark drip-dry and Christmas perfume. My Lord Telfer looked well, methought—a trifle pale perhaps, but not surprising, considering the cost of the dinner.

The long tables in the banquet hall being already set with a salad composed of many varieties of leaf, two kinds of dressing, and crackers, there entered promptly eight lords and ladies in old English costume, singing songs of welcome in Latin, and betook themselves to a long table on the stage, there to sing and chat gaily. One of the lords wore a full-gathered beret low on his brow like Beetle Bailey.

Shortly thereafter the trumpets sounded from the balcony and there entered two servingmen in white tights (to which feet were attached as on a child's sleeping garment, but grown up). They wore long, narrow panels on which, methought, a display of advertising might have been placed, and thus the dinner cost might have been alleviated.

Between them they bore the stretcher on which, among boughs of evergreen, rested the punch bowl, filled with punch of a seemly red hue, though a trifle insipid of flavor, if one might judge by the cups already placed before us, in which we now drank a toast.

Then blared again the trumpets, followed more madrigals, and the company on the stage sang the Boar's Head carol, oldest printed carol, written in 1526 by Wynkyn de Worde.

And now, on the evergreen-bedecked stretcher the boar's head was borne in: brown, greasy, its ears upheld as if by shirt-collar staves; a red-and-green-striped apple held open its mouth, disclosing tongue and teeth; glazed red cherries filled its eye sockets. Gazing on this splendid thing, I did remind myself that early on the morrow I must get the lard cans ready for Poor Richard to take with the meat hogs to the freezer plant.

Richard, who hath wondrous understanding in these matters, glanced at the passing boar's head and remarked: "A small Hampshire hog, about two hundred fifty pounds, with nice, refined nose."

The plates now set before us bore beef instead of pork; also Yorkshire pudding, Brussel sprouts, browned potatoes. Hot rolls were brought, without trumpet sound. With the beef we were served that most excellent brew, coffee, dark and bitter, poured from small shining pots carried to the different tables by charming young ladies.

Last the flaming plum pudding was brought in on the green boughs, with proud trumpet sound. Small fluted puddings being already placed before us, we ate them while a harpist and a large company of graceful young women entered, singing as they strolled. They sang old Latin songs to the thin music of muted harpstrings, enchanting to the last dying-away vibration. The company then dispersed. At Fair Dodhead Farm we stopped for a more robust toast and so home. And God rest ye merrie, my Lord and Lady Telfer.

❁ THIS MORNING the sky was pink in the east, like a cloth that has been dipped in pokeberry juice and hung up to dry. The corner pond and the little stream running into it were pink, too, from the reflection, as if a little pool of juice had dripped out of the pink cloth and started to trickle away.

❁ THIS LAST DAY of December Dick brought in a square flake of hay baled last summer. It was from the second cutting, in mid-July: long, wide, pliable blades of orchard grass and stems of alfalfa and clover, still green, with its flat-pressed blossoms still pink and resilient. The hay gave off the sweet, fresh hay odor one associates with clean, old barns.

"Feel it," he urged with pride, "how springy and undried-out it is. It's more like fabric than dried vegetation." I felt it.

"Pinch it," he insisted. I pinched it.

"Take a whiff of it," he said, holding it to my nose while bits of it fell out and lay on the back step. "Don't it smell wonderful?"

I took a whiff and it did smell wonderful, and the year wheeled suddenly back to mid-July. I had a quick glimpse of the low, rubber-tired wagon, stacked high as a corncrib with the long, square-cornered bales; and on top of the rocking load sat Rose, the two farm children, and a pretty aunt, Katsy, who was on vacation from a job in Munich. That day the summer had been in its prime. Nothing holds time more unchanged than a fragrance, or a wisp of music.

Dick laid down the hay and picked up a square of baled oats straw, coarse, flattened from the pressure of baling, clean yellow straw with the grain still in the heads. It smelled pleasant enough to sleep on.

"And you can grow that from start, on almost any farm," said Dick, who has used it to put some meat on the starved bones of this farm since we came here.

He started away, picking up both hay and straw, then turned back abruptly. "You break open a bale of hay and farmers call it a 'flake' of hay," he exclaimed annoyed; "the county agent calls it a 'fleck,' and it just don't sound right. A fleck is like a fleck of dust on your coat sleeve, a fleck of foam on a horse's mouth. It just isn't the right word for hay."

JANUARY

[JANUARY]

JANUARY came in suddenly one cold wet day this week, and at midmorning I heard the sound of grateful footsteps seeking refuge in the back room. When I looked in, Dick was sitting there by the little drum stove, with all his outside wraps still on. "This is the place to do your farming today," he told me cheerfully.

He opened his desk and took out cigar boxes filled with papers and receipts. "I can work on the state gross," he said, and his look of gratitude would have been a real shock to the collector of internal revenue.

THE LONG, stiff card that comes with the state gross income tax blank, "to insure proper credit" for your return, is perforated by numerous slits in a mysterious arrangement that makes it, actually, a piece of a machine. By this card, moving routinely through a machine like a wad of hay through the chamber of a hay baler, certain facts about you can be picked out, assembled, reclassified, and analyzed.

The perforated card is as individually an expression of you as your monogram is, or your signature. Its slits cannot add up to anyone but you.

With this flattering fact in mind, the state gross returner is tempted to use the card as a stencil for marking his personal garments or household linens.

JANUARY has bleakness; naked trees with cold rain dripping from them. It has drabness; violence; sharp, bitter winds; and the colored magnificence of ice, too. But sometimes it has moments of tenderness that are not surpassed by the compassion of any month.

One of these is that moment when night has just barely come but day has not altogether departed from snow-whitened hills and fields. When lights come on in the farmhouse then, one looks from a warm, lighted room into an outside world turned suddenly bright blue.

Against the glasslike clarity of that blue sky, bare trees loom up in blackness. A line of fence posts rises up blackly from fluffy cascades of black weed and brush. The last, late birds, hurrying from feed pans to wherever they go to sleep for the night, are small black pebbles tossed against the bright blue air.

This moment forgives the day all its shortcomings.

This late, bright blue moment of early evening is reward for one's having ploughed laboriously through snow all day, to the barn, mailbox, henhouse, to neighbor's houses, to woodpile or cistern or wherever one went that day. The blue world lasts only about ten minutes, fifteen at most; even the busiest of farmers can spare the time to accept this tenderness from January's cold hand.

WHAT DICK really wanted was not just fried pickled pork, but fried pickled pork that tasted just as it had to a hungry little town boy visiting his Aunt Belle and Uncle John on the farm near Jordan Village.

One of our earliest butchering efforts, therefore, concerned pickled pork. Following instructions in a government bulletin, we put strips of fat, white pork into strong brine in a ten-gallon jar. It was not successful. The brine became ropy. By the time the meat should have been ready to fry, it was ready to throw out, which we did.

Dick's yearning persisted for years, and last week it broke out in church. There he made arrangements with our neighbors John and Leota Dunning to pickle some of the fat sides from the hogs Dick was taking to a commercial locker plant for butchering.

Butchering, for the most part, is not a farm-done operation any more. Some farmers do cure and smoke hams, bacon slabs, and shoulders, but most of them have the whole job, including the cutting up and wrapping, done at a custom-butchering place. Pickled pork is not included, as routine.

In the evening John and Leota came, she wearing a big apron and carrying a long wooden spoon. I got a new box of table salt down from the kitchen cabinet shelf.

"It takes a good deal of salt," said Leota tactfully. "For a jar this size, about fifty pounds I 'spect." Dick went up to the barn and brought down a sack of stock salt, and when we had used all of that he and John went to John's barn and got another sack.

In the bottom of the jar we laid two small pieces of wood, to prevent the brine from holding the meat in a relentless suction at the bottom of the jar. "Anything except pine or oak will do," said Leota. "Pine tastes like tanbark and oak tastes like something I wouldn't even mention." We used maple. On top of the wood we laid the first strips of fat, about three inches wide and two inches thick, with the shaved pigskin left on. We packed it all around with coarse white salt, repeating this until the ten-gallon jar was full.

While we worked, Leota told me how to cook the pickled meat. "Cut off the rind, slice the meat, and soak it overnight. If you're in a hurry, soak it in warm water; but this spoils the crispness. Put a little grease in the skillet to keep it from sticking. Dip the slices in flour, and keep turning them in the skillet. Fry out all the fat; fry until it is crisp and brown, like mush, and kinda blistery-looking."

When the last layer was surrounded with salt, we made brine strong enough to float an egg (testing it with an egg), and poured this over the meat until it was all covered. Joe went out and brought in a big flat stone, which I scrubbed, and we laid this on top of the meat. Then John and Dick carried the jar out to the freezer shed, and we covered it with a clean cloth and newspapers. In six weeks, Leota said, it should be ready to sample. We wiped away the salt from the kitchen table and all sat down and ate Christmas fruitcake and ice cream. The children went to bed, but the rest of us stayed up late for the visiting that is a farmer's reward for work.

"THEY'RE all right," replied Dick, pulling on a new pair of gauntleted, leather work gloves Santa Claus had brought him, "except there's no place on them to wipe your nose."

UNLESS a bitter wind is blowing, or rain is beating in through the empty windows, the barn is a snug place on a winter morning. When Dick called me up there this morning, the barn was full of a dusty atmosphere of abundance and content. The cattle are shaggy-coated and fat. The hay they were eating had a pleasant odor, reminiscent of summer. When they opened their mouths for a refill, their fragrant breath came out in a crooked, wafted column, like steam. The winter calves are dark red, too, and sturdy. Their blocky, triangular heads are positively winsome, and they were not afraid of me when I reached out to pet them.

In the stall next to them, a yearling bull calf was sitting up on his haunches, like a shaggy red bear. (That was what Dick had called me up to see, for the position is unusual.) While we watched him in amusement, he went on sitting, placidly chewing, and looked back at us unconcernedly, as any native looks at an amused tourist.

On the edge of the top board, by the bull's stall, a row of Bantams were comfortably perched, all facing the same way. They were small, brightly-colored, and picturesque, like something out of a storybook. The one nearest the wall was the most graceful and gaily-frocked. He had long bright feathers on his yellow legs. "That's the keep-rooster," Dick said. "I'm going to sell the other five this week."

CAROL READ the health lesson aloud while I washed dishes in the kitchen, and thus we both learned that the human body is 90 per cent water. To me this sounded like understatement. There are times when a farmwife's body can believe that she is 90 per cent dishwater alone. Add another 5 per cent for laundry, some for washing potatoes and fruits before cooking, and a panful for scrubbing the linoleums in muddy winter weather, and it wouldn't be surprising if the total came out somewhere nearer 135 per cent.

The health lesson continued with the information that a number of familiar elements of everyday farm life are also largely water. An apple, for example, is 84 per cent, a grape 77 per cent, an egg more than 73 per cent.

With such small amounts of solid in these different

items, you'd expect people, apples, grapes, and eggs to look more alike. The human body, being only 6 per cent removed from applehood, looks surprisingly unlike an apple, except in the cheeks. The cheeks of healthy children sometimes look like the cheeks of healthy ripe apples, especially in winter weather, and the cheeks of old people sometimes resemble the cheeks of old apples; but nobody has much genuine difficulty distinguishing between these two forms of largely-water content.

The character of that small per cent of solid must therefore be remarkably stabilized and individual.

There were some provocative facts the health lesson didn't explain. For example, what per cent of the human spirit is composed of weather? What per cent of the spirit of an apple, egg, or grape is made up of weather?

Even before the human eye consults the big, highly readable thermometer on the outside wall of the house, the rising or dropping of mercury has already had a marked effect on the level of the human spirit. If you observe the spirit in both autumn and spring and note the different moods brought about by different kinds of rain or wind, you come inevitably to the conclusion that the weather content of the human spirit is at least as high as the water content of the human body.

The quality of solids in the different spirits also varies greatly. The human spirit and the spirit of an egg will both freeze in sub-zero weather. The egg will swell and crack; the human spirit is more likely to shrink. Both will thaw, but with a difference. When the egg thaws, it unswells, its crack closes, and it looks all right; but its yolk is stiff thereafter, as if somebody had started to boil it and never finished. A thawed egg spirit isn't much good thereafter, but the thawed human spirit is as good as ever.

THE TWO little Bantams, hatched late in November by a Bantam mother who stole out her nest with characteristic Bantam independence, have now feathered out completely. Their survival is a contradiction of all the modern rules for poultry success.

Early in December, when the first snow came, I saw the little Bantam mother standing in the driveway, halfway up to the barn, in snow up to her first leg joint, where the feathers set on. The two downy chicks were perched on her back and cheeping persistently, unhappily; but the mother simply stood there, unable to decide whether to make the super-Bantam effort of going on, or to wait there, throwing all the responsibility on the Lord. The Lord came to her rescue, in the person of a farmer.

There were many cold nights when she abandoned her two chicks and went to roost with her contemporaries in the henhouse, leaving the chicks yapping outside. At such times it was necessary for the Lord to send someone to pity the featherlings and put them under their mother's wings. She was willing to accept them.

They never had a warmed brooder house with temperature beginning at 95 and being gradually decreased as they grew. They never had scientifically balanced starting mash, cautiously overlapped with growing mash at the right time, to avoid throwing them into a premature moult. They never had any safeguard, except their mother's native resources, against predatory nocturnal enemies. They followed their mother through rain, dew, or sunny weather, and ate what she pointed out to them.

The heavier, cultivated breeds of chicken, if thus carelessly brought up, would have been dead and out of their troubles long ago. There's a philosophy there, if anybody has time or inclination to follow where it leads. It could lead off into speculation about such things as hunger and survival, independence and comfort. It could lead to Washington, to a bulletin about better farming methods, or poultry care. It could lead down to the henhouse. Or it could lead out of the chicken business altogether.

SNOW BEGAN in the dark morning, with big soft flakes tumbling in a newslike urgency from a dove-gray sky into earth's brown lap.

There are things that can be compared to snow; colored leaves falling can be, or ash from an autumn bonfire, or confetti from a hero's parade. But snow cannot be compared to anything, because it is not really like anything else.

Near the house the whirling snow squares were caught in still air and slowed down, but out in the clover field the tumult of their falling inspired the fleeing sparrows to greater speed. The birds fled past the kitchen window to the shelter of the althea bush where the bird feeders are. Dish towels on the west clothesline tried hard to get to the shelter of the front porch, while steadily, on the road outlining the farm and on cattle paths in the fields, the white deepened.

"Why do people always get so still when they watch the snow," murmured Carol at the kitchen window, "as if they could see it better that way?"

It is because the beginning of snow is, itself, a profound stillness in which the listener hears the thoughts of his own secret heart.

SNOW ICE CREAM was a country luxury in the old-fashioned days, and it still is, because any place else now, the snow would be suspect. Someone would recite statistics concerning radioactivity.

On an Indiana farm, though, and some other places, you can still make snow ice cream when the snow is just right. It must be freshly fallen, fluffy free of any suggestion of earthiness or smog.

Go out to the cleanest drift in the yard and bring in a big panful of snow. Add sugar. Add vanilla. Add thick yellow country cream, freshly skimmed from a gallon crock of milk. If you are desperate enough, you can use half and half, or commercial coffee cream, but in that case don't expect to get bona fide old-fashioned snow ice cream.

Mix all this together lightly, lightly, lightly—in just about that many quick, folding-in strokes.

Serve immediately. And don't dawdle over eating, either, unless you like to drink ice cream. If the telephone rings, pretend you didn't hear it. If you eat too fast, or take too-big, greedy bites and get a sudden sharp stab of headache in the upper corner of your forehead, a bite of cracker will stop it. Veteran eaters serve crackers for that purpose with snow ice cream—just plain, salty, square soup crackers.

THE WAY of deep snow with sound is extraordinary. Right now, when the roads are filled with snow, there is a remarkable difference in the way the larger, common sounds come through the snowy air to the snowbound farm. (Sounds, in fact, are about all that can come through right now.)

This morning the diesel's horn, which ordinarily sounds like a petulant cow, comes through as an almost pure organ tone. The chattering of English sparrows, struggling for status at the ground-corn pans, has become the crisp sound of castanets.

At night, the sounds are skeletal, like the veined network of a leaf after the rest of it has disintegrated. From close to the house you may hear the small crying-out in sleep of a winter-lingering bird; but now the woven-fabric sound, made in summer by the humming of many insects, is all gone. The winter sounds, coming in through closed doors and windows and cold walls of a farmhouse are sibilant, vestigial. One hears the dry squeak of wheels or feet on snow; the sound of a cow bawling at the barn, or a train passing in the night, or even of a rooster crowing in the night, is stripped thin.

Now the night listener, lying awake, hears the flame flapping softly in a stove in the kitchen, the whirring of an electric motor at furnace or refrigerator, but, of outside sound, practically nothing except the muttered argument between the old farmhouse and the wind.

"DO THE winter fields look pretty to you?" asked Dick as we stood at the front window watching the children get on the school bus. "I thought the view of bare trees and snowy fields, through John Fielder's living-room window yesterday, was just simply beautiful." John's window looks out on a faded red barn, a nice-sized creek, and a stony hillside with two dead elms at the foot of it. "Actually, if the rain's not running down your neck, and your feet aren't cold, any winter day on the farm is beautiful," he added.

Snow changes the colors of a farmscape. Against the white of new snow the cardinals look more acutely red than at other times. Even the female cardinal, whose red is limited to a deep, exciting rosiness, blends it with snow-grayed brown, adding depth. The high, ceramic blue of blue jays is much diminished when it has to compete with the sharper, harder blue of snow, but the black chickadee becomes dark blue.

THERE IS something at once magnificent and terrible about moonlight on the snow during these cold, white January nights. The burning moon consumes the darkness, leaving black bones fallen from every tree and fence post. Footprints on the ground are tangled shadows; their dark trails lead restlessly from the road, from the house, from barns and fields, across the stark snow.

If one is outside, and cold, the landscape has a profoundly lonely look, underlined by shadows. Only in a warm house, surrounded by people and the pattern of busy living, is this loneliness abated. Even then it is awesome and beautiful.

By moonlight the cattle come down in slow dignity to get a drink at the corner pond. They walk on the snow-covered field as casually as if it were summer and the grass thick underfoot. Caught between brilliant snow whiteness and the moonlit brilliance of winter air, the red and roan cattle are all turned to black. In contrast to the snow's white, Jupiter, the white Shorthorn bull, has become a silky silver-gray. He stops to rub against a small bush near the fence, and one is startled to note how moonlight has transformed the bush, burning but not consuming it, like the burning bush seen by Moses for a sign.

DICK LOOKED through the mail and laid aside two cards that were obviously birthday cards.

"Happy birthday!" he said, sardonically. It was his birthday, but not happy. For some reason every year this day seemed marked by unusual misfortune. One year the tractor broke down in the field with a load of manure on the spreader; another year a cow sickened. It turned out to be forage poisoning, but the veterinarian suspected rabies, so the cow had to die slowly and Dick had to take rabies shots.

This year the best Shorthorn calf was sick.

All day we had been unable to get a veterinarian, so we had doctored the calf as one would a sick baby. The calf was co-operative, the mother standing near seemed to realize that we were trying to help. Between times of pain the calf stood up and pressed his little square nose against her side as if he thought she might help him, and she licked him tenderly with her long, rough tongue.

By the time we finally located a veterinarian who promised to come that evening, the calf lay exhausted on the straw. The rainy witchlike day had worn down to evening. The path from the house to the barn was a slippery streak of mud from much going back and forth for soda, warm water, mineral oil, turpentine, telephone messages.

At the house the wrappings of birthday presents, opened at breakfast so the whole family could share in the pleasure, still lay on the floor. Bedtime came on in an atmosphere of gloom. When Dick came in from the barn, everybody silently searched his face for news.

"I can't see as he's any worse. He tried to stand up just now, but he's weak."

Dick is a good cattleman, acutely observant and keenly aware of the different personalities of his animals. He faithfully reads the Shorthorn sales catalogues and breed magazines and knows the history of the good herds as well as his own family's history.

His father and grandfather were livestock farmers,

and his grandfather was also a banker. His father went to De Pauw University a year but considered it a waste of good farming time. His mother went to normal school and taught before she was married, so perhaps it was natural for Dick to feel that he ought to teach although he wanted to farm.

He had taught in a country school one week before the township trustee yielded to the county superintendent's opinion that a year of general agriculture at Purdue did not qualify a young man to teach school. The young teacher was therefore reluctantly dismissed, the school closed, and its eight pupils hauled away to another school. There was one little first grader who must have suffered from the change. She was barefooted and shy, wore her taffy-colored hair in two tight little braids tied at the ends with grocery twine, and she could read only when she stood by the teacher's chair, with his arm around her.

It was just as well for Dick that he had to give up teaching. His farmer father had always lived in a small town, but Dick's heart was set on being a farmer and living on his farm. Of the four sons in the family, he eventually became the farmer. A year at Purdue University did not change his farm way of talking, any more than a comb could take the curl out of his dark hair. But a man can be poetic by nature and yet express himself in the handy, accurate language of farmers. It might even have been one of the ways he unconsciously expressed his yearning for a farm.

He has prejudices and strong personal opinions. He is left-handed, tall, slightly stooped, and now wears bifocal glasses. He likes to read; prefers the sophisticated *New*

Yorker magazine, which he has read regularly for many years, to the farm magazines, which he finds depressing or irritating. He is acutely observant. He is a very present help in trouble, and sometimes a very present trouble, needing help, as on that birthday when the calf was sick.

We were at the barn when the veterinarian finally came. He gave the calf a shot and a dose of medicine; said to give it another dose after half an hour and another two hours later, which would be around midnight.

"We may as well stay up here until it's time for the next dose," I suggested. Dick went around in the barn finishing the chores that had been delayed by the calf's illness. He poured ground feed into the big center trough. The cattle ate eagerly, pushing each other aside and getting the powder into their nostrils and licking it out. Dick opened a bale of hay and I put some under the sick calf. The cows pulled out good bites of the hay and chewed it, moving their underjaws slowly from side to side.

The light bulb was dusty, dimming the light. A layer of dust and chaff lay on the wood surfaces, but where the cattle rubbed the stanchions with their necks, reaching the hay, the boards were glassy-smooth. A barn accumulates a strange overflow of obsolete furniture from the house. I recognized the narrow book cabinet that now holds boxes of nails, fly spray, small tools, and cans of salve as the one abandoned from the house several years ago.

I sat on a bale of hay and watched Dick bring in a bucket of water from the faucet by the door, and water the bull. Then he came and sat down on the bale beside me and we talked. The cows breathed deeply into the hay. "Cows' breath is one of the pleasant smells of a barn, isn't it?" I remarked, having noticed some less pleasant ones.

"Yes, but do you know that when a cow is angry or frightened its breath has a different smell? I've noticed it about the bull." He paused, frowned slightly, concentrating. "You know how it would smell if you went out in the garden in spring and a bale of hay had been left there all winter and you took a fork and broke it open?"

I did.

"Not really unpleasant, you know; just sort of musty, like old rotting straw. If you ever smell it, you recognize it as the smell of fear or anger."

We gave the calf the second dose and tried to believe he seemed stronger.

"I'll set the alarm and come back at twelve o'clock," said Dick. When we turned off the light, the compassionate darkness came in. The young red pigs were nursing, making a smacking, foamy sound.

"I'll come back with you," I said, as we went down the muddy lane. Partly from a sense of loyalty, partly because I like being in the barn at night, mostly because I hoped the calf would be better because that would be the best of birthday gifts. (As, fortunately, it turned out to be.)

MY NEIGHBOR Fred Dutton today expressed one of life's really great truths: "When the children are all right, everything's all right."

How is it in Helen's woods now, I wondered. For along the road up to Helen's house, the bark on young sassafras bushes was bright green, showing why Indians called this aromatic tree "green stick."

The snow was softening, but the gray sky promised more soon, as Helen and I walked back into the woods behind the Lewis's house. It has not been cut over nor pastured for many years, and is whiskbroom thick with trees that are tall but not very large in diameter and have few limbs because they have had to reach so far for a sip of sunlight.

It is not necessary to go deeply into a woods to see how relentlessly winter breaks the faltering, uncertain grip on life. The woods floor was strewn with the limbs and whole trunks of trees that were dying last year. Brown and oyster-colored shelf fungus, bright green lichens, and the harsh green plush of bright moss had made good use of the deadwood. Under the melting snow, wet brown leaves lay like layers of brown wrapping paper. Where Helen and I stepped, pressing the snow into transparent glasslike thinness, the brown gave our footprints the pale green color of old glass fruit jars.

It was warmer in the woods than it had been along the road. We felt no wind, although the tops of tall trees were swaying gently. Among their branches we saw many abandoned nests of birds and squirrels. On tall shagbark hickories the inelastic bark had split lengthwise, to accommodate the increasing trunk girth. The bark hung in ragged strips, "like old clapboards," Helen exclaimed.

There were many tall beeches, their smooth bark unbroken and gray. There were oaks that did not look like oaks, having had no room to throw out craggy, outspread arms. We identified them by their brown leaves, some with needle-sharp lobes, some delicately scalloped in an elongated, tapering oval, some with deeply rounded sinuses. The leaves of oak trees are a literature unto themselves.

"Aunt Beulah found ginseng in the woods here last year," said Helen, "but I don't know just where."

There were many tall, old sassafras trees fourteen to eighteen inches in diameter. They were all crooked, as if from having had to peer first over one shoulder and then another, never being able to stand up straight at all. Their dry, faintly aromatic bark was bright brown. Where it had been broken away from the trunk, as at an inverted V in the base of one where a squirrel had an entry, or higher, where a red-capped flicker was pecking out a saucer-sized door, the bark is orange-colored. The trees were old; their bark had achieved a woven braidlike pattern, deeply sculptured, which is proof of age in the "green stick."

EARLY this morning I took a field glass and watched a rabbit sipping snow from weeds in the raspberry patch, and I am convinced he knew he was being watched. As soon as I had got the glass focused on him, he suddenly became very still and then sat up on his haunches with his side turned to me so he could stare intently back. I could see the dark, oily shine of his eye, the gentle moving of coarse brown and white fur as he breathed.

Presently he let himself down on all four feet and resumed his sipping, running in little circles, hopping and hiding as if trying to keep a screen between himself and the eye he felt on him.

Perhaps, being wild and keenly aware of danger, he always carries on in this guarded way. But I have seen birds become restless under a steady human gaze, and I have seen Rose avert her eyes, as if it were painful to be too steadily looked at; and I believe animals have a sixth sense, which is the sense of being watched.

People have it, too, to some extent. Everyone knows you can often get a person's attention by staring steadily at him. And children know that in playing hide and seek, you must not look at the person seeking you, because he will see you if you do.

IN SUMMER when all the leaves are green you hardly notice it, but now along the hilly, country roads you see patches of bright, toast-brown broom sage.

It is not a popular nor a tended plant, but it has its moments of importance. One of them is now, when January fields have the drabness of a child's colored sock that accidentally went through the bleach water.

From a distance a patch of this tall, harsh-looking, grassy plant resembles a giant's shoebrush. It grows where little else is growing, on cutover, untilled, hungry land on steep slopes. This is its first merit. When winter rains threaten to gnaw the hills away down to their bare bones, the tenacious roots of broom sage hold the soil there. One of its cousins, Hungarian or smooth brome, is used for holding creek banks in place. Another cousin is used in the south for forage; another, called "chess" or "cheat," is an outlaw and poses as wheat, to plague farmers. Another, called "ripgut," is injurious to livestock that eat it.

Dick always has a good word to say for broom sage because, he says, if you mow off the first growth, it will come up again from the root, without a harsh stalk, tender and palatable for livestock.

THE SPOTTED pony mare had a filly colt three weeks ago. No bigger than a big dog, coated in long-haired winter fur, the little colt looked somewhat like a winsome donkey. She died after four days, although the veterinarian who gave her three shots for pneumonia thought she would pull through.

We thought so, too, having brought her into the house to a bed on the floor in the back room, made of the finest fragrant alfalfa hay, and having fed her milk with cod-liver oil and corn syrup.

She was responsive, grateful, endearing. She had beautiful green eyes from which the deep jewellike light departed the abrupt instant of her last breath.

A common fact, but miraculous still, is the fact that the colt born into a winter world had, during her prenatal life, been furred for winter living. A spring colt would have been born with short hair. It is revealing of the way nature communicates necessary knowledge to new creatures, hiddenly developing within the warm insulation of the mother's body. It reaffirms something that is neither explored nor altered, nor wholly explained, by science. It also tosses out the window and tramples into the ground the theory that science and religion are incompatible. The further science goes into commonplace miracles, the more miraculous science itself becomes.

THE ICE SHOW began on Friday, and for three days we lived in a glass world.

Early Friday morning sleet was falling in a rasping, hurried whisper against walks and windows and into treetops. By daylight you could see what had already been accomplished. The bent-over peach tree was a glass tree; the wire fences, a fat crisscrossing of glass lines. Long pliant maple branches, dotted with reddish brown buds, were glass-dipped and swung softly as if about to break into tinkling music. The whole farmscape was a wonderland of glass, wet and shining. When the school bus came, cautiously, the children went down through a glass yard, under glass trees, down to the glass road. Trying too hard to walk carefully, Carol fell once. Joe abandoned himself to the slippery mood of the show and fared better.

The next day the sun came out, making the ice dramatic. Now the stiff trees turned to silver. The still air had the shine of silver from earth to sky, and every little insignificant weed and bush had been silver-plated. The silver trees seemed even taller than the glass ones had been, and more burdened; and that day there was an occasional muffled crash as a silver top broke and fell down.

By noon silver stars burned from every limb. The clove bush beside the gasoline barrel stood in unheard-of silver splendor. Its branches had become pure, fragile silver, and all over it the sun hung stars burning in Christmas colors—red, copper, yellow, blue, green.

On the third day there was no sun. The brilliance dimmed to a gleam. Trees glowed softly under a dull, blue-gray sky. Many were enclosed, separately, by a strange, pale-gray halo. In the garden, old weeds were transformed into lace, spun delicately like foam. Driving along the narrow country road from Herschel's lane, we looked down into a cavernous ravine filled with silver forests. The car was occasionally pommeled by an armload of silver splinters from trees overloaded and leaning against each other's shoulders across the road.

That night the moonlight was clear and deep on the crusty snow. Trees stood white and silent, as if they knew what was to happen. Then rain came, melting the glass and silver burdens. By morning the splendor of the ice show was gone. It had been spectacular, but we were glad to have the world back in its customary dress. Brevity, soul of wit, is also the boundary of splendor.

"A FELLER just lives as long as he's supposed to and no longer," said my eighty-seven-year-old neighbor Eaglie

Stanger. She had walked the quarter of a mile from her low, stretched-out old farmhouse up to her mailbox. The mail wasn't there, so she came on around the curve and up to our house.

"Are you home, Rachel?" she called, tapping on the window pane. Her voice quavered slightly, just as it does when she sings in church.

The special charm of Eaglie is her almost childlike pleasure in life and people, what they do and say, their houses and children, and food, and the gifts they bring her, which she opens immediately like a happy child.

Her eyes are bright and blue-gray, and she wears glasses only when she reads. She has a pert, short nose, wears her silky gray, curling hair in an upsweep from her neck and ears (as she has worn it for seventy years), and keeps it in place with gray side combs. Her face is wrinkled like a potato in late March.

She had on a printed apron over her cotton dress, and a sweater, and carried a long, crooked stick polished from having been carried on so many of her walks, and before that by her husband Clint. Eaglie used to say: "Oh, Clint is the best old man! The only thing I love more than Clint is tea," and then she would laugh. She lives in the house she and Clint lived in for more than fifty years. His parents were living there when Clint and Eaglie moved in. Her son Carr and his wife Iris have lived there since they were married.

"I've had some hard times and worked awful hard in my life, but I always enjoyed life," said Eaglie, and chuckled. "I love company and I've had as much company in my time as anybody. I'd like to have one more big dinner." She began to name the people she would like to invite, and I knew that her walnut drop-leaf table, even with all its extra leaves inserted, would never hold half of them.

In summer Eaglie always has flowers in the long flower box out in her yard, and I always send her a red flower. That morning I gave her a little red coleus I had been saving for her.

When we moved to this farm, she was one of our first guests. She brought us a fresh pumpkin pie and spent a happy hour telling me the history of this house and farm and of the community.

At that time Joe was two years old, a friendly, talkative, yellow-haired baby. Almost immediately he and Eaglie discovered the lovely understanding that exists between first and third generations. On Sundays it was his self-appointed responsibility to take a gift to church for Eaglie. Apples, when we had apples; potatoes once, when he could find nothing else. She accepted them graciously, and when she cooked them, later, she and Clint had a tender, good laugh over them.

Eaglie gave me a special gift once. She had tacked up a handmade quilt against the inside back wall of her brooder house to keep the draft away from her baby chicks. I protested about this quilt abuse so much that once she exclaimed: "If I had that quilt washed, I'd give it to Rachel!" The next time I went down it was hanging, clean and dry, on the clothesline in her yard, and she took it down and gave it to me.

The first time I visited Eaglie's house, I went with my nearest neighbor, Rena Dutton. We were standing in Eaglie's bedroom, a small room with painted white walls, blue and white curtains, a favorite pink geranium at the window. There was an old-fashioned cherry bureau with a swinging mirror there, and on it an old white vase shaped like a hand holding a flower container; there was a Jenny Lind bed and a little footstool to step on to get into it. There was a small square mirror on the wall, and hanging under the mirror one of Eaglie's several comb cases. There were two rocking chairs, one for Eaglie, one for Clint. She said: "Rennie, I'm awful proud of my home. It ain't much, but I'm awful proud of it." "Proud" is her way of saying: "I love it dearly."

Much as she loves this world and her neighbors, the church and the hymns, the vegetable garden and flowers, the silo dinners and church homecomings and the old house itself, Eaglie has all the essentials, except stockings, folded away in her bureau drawer against the time she leaves this world.

That morning she drank tea while I drank coffee. She sat on a straight chair by the window so she could see the mailman when he stopped at her box. When she saw him, she stood up and retied her scarf. At the back step she put on her overshoes. "I have plenty a'ready for dinner," she said, "but I want to get home in time to bake a pie." I offered to take her home, but she refused vigorously. "I like to walk and it's good for me," she explained, picking up her stick. I walked down as far as my own mailbox with her and there we parted.

"Now, come down, Rachel!"

"Oh, I will, Eaglie. And you come back."

Eaglie is the pet, the tyrant, the matriarch, the special flavor, of this community. I love her dearly.

ON THE snowy path up to the barn this morning I read again the unintentional autographs of several small creatures that share this farm with us. In the thin snow on the back steps and below them were the scanty, scriptlike tracks of birds. They are shaped like a Y, with three prongs in front, one at the back. There were several prints around the white-rose bush. Under the peach tree straight parallel lines of tracks met at right angles, beginning and ending suddenly. When the going got too difficult on foot, the bird had simply taken to the air. I envied him.

There is a difference in bird footprints depending on whether the bird hops, leaving parallel lines of quotation marks, or, like a quail, runs, leaving a single line of prints. Where the quail crossed through deep snow toward the ground-corn feeder, his feet sank deeply, making a diamond-shaped track.

At the edge of a road we saw the sharp scratches made by a crow's wing as he took flight from a low limb and his long feathers brushed the snow.

Under the maple the bird prints looked as if the bird had been wearing skis. "That's that long back claw," said Dick. "If you come up to the barn, I can show you where two foxes caught a duck last night. You can tell which fox carried the duck. He waddled a little, feet wide apart like a mother cat carrying a kitten in her mouth."

Visiting dogs had run back and forth toward the barn. The deep cups of their tracks showed little, pressed, round prints with four toes and a larger pad like the inside of a cookie mould. A cat had traveled the path, too, but more lightly, on smaller, daintier feet. Once in a while she had deliberately made clusters of footprints, as if trying to hide her intentions, or had stopped to do a gay little dance in the snow.

Near the corncrib a hurrying field mouse, sinking deeper than it expected into the fluffy new snow, had left a short double line of tracks, almost like links in a chain.

A rabbit had waited under a tree near the house. His small front feet had been brought up close together, making one deep, large hole; the hind feet had landed in front of them, leaving a couple of long quotation marks. "You can see he rolled a little there and threw snow beyond as he caught his balance," Dick said.

We watched a skunk run up the short, snowy slope toward the shelter of the church fence. Behind him he left a marvelous pattern: straight lines composed of four diagonally spaced dots, as even as if measured with a tape. A squirrel, also, had come down from a tree, landing deeply in snow, crossed a few steps, then resumed his tree-traveling again.

The birds' autographs were my favorites. With a pancake turner and great care, I managed to get up one whole unbroken autograph and put it into the freezer, to read again, I hope, next July.

FEBRUARY

[FEBRUARY]

GROUND HOGS cleaned house this week," reported the chairman of the first-with-the-news committee. "I never knew them to push the dirt out from around the tops of the den in February before."

IN FEBRUARY a farm is like a child's hand, opened and held palm up to show that nothing is concealed in it. Without leaves to veil the farm buildings and fields, the naked facts of a farmer's work habits and holdings are revealed even to the casual traveler passing along the road.

Did the farmer get the disk put away after he finished

the wheat field last fall, or is it still out in the field rusting away its spring trade-in value? Are the cattle fat or thin? Are the hogs fenced in well and nose-ringed or are they out rooting up the tender sod of the pastures?

Nothing is hidden in February.

On the red barn wall is the basketball goal, essential part of Indiana farming equipment where there is a teen-aged boy. It is surrounded by muddy blurs, thumbprints of a ball that missed its aim but never lost its hope.

Now the farm's colors are austere shades of gray, brown, black. Yet, within the limitations of these drabnesses, nature achieves a variety of dramatic statement, wry comment, and rich mood. Rosy-brown maple buds, in boutonniere clusters, contribute gaiety. The yellow-brown of dead grass barely conceals the impatient, short growth pushing up at the crowns. Where the hoofs of hogs and cattle tear the sod, the fields are a rich, wet chocolate-brown color.

"And remember," said Dick, "Uncle Howard always said that in February it thaws a little every day."

THERE IS a rabbit that comes into the strawberry patch now where the brushy young volunteer plum trees are. We have spent some interesting moments looking at each other, the rabbit and I. I sit in the warm back room; the rabbit sits on the snow; and we both enjoy the plum trees, but for different reasons and at different times. I like them for their bloom, which is sweetly fragrant but never comes to much in the way of fruit because the trees are never properly sprayed. The rabbit enjoys them for their bark.

Today, a cold, bright, Currier and Ives morning, sunlight warms the graying boards of barn walls and brings a shine to the young bark of the plum saplings. The rabbit has been out, nibbling the bark, which exposes the white inner wood, causing it to turn a weepy, pale brown like a bitten apple. The plum tree will survive this injury only because it is a volunteer (as farmers call a tree that comes up, unplanted) and these are usually more rugged than budded expensive ones, which would have to be burlap-wrapped for protection against the ravages of the rabbit.

By going only twenty hops farther the rabbit could have had all the rich yellow corn and fragrant alfalfa hay she could encompass. Surely the dark and bitter taste of plum bark must be a stimulating delicacy to her, like the cup of black coffee I hold, sipping luxuriously, while I watch her.

THE SUN was bright, but the cold was piercing, as I walked up toward the big pond back of the barn, to investigate the strange, wonderful ways of water with earth.

In the doorway of the barn a wheelbarrow, freshly filled with silage, was standing. Steam was rising up from it into the cold air, like steam from a gigantic kettle of soup just taken off the stove. At this time of winter, silage still has a pleasant pickledy smell, with a suggestion of sorghum. Cattle like it. The cows in the barn could smell the steam and were bawling in happy anticipation.

"You can actually get your feet warm, up in the silo," Dick said, and came out to go with me to the pond.

The night's cold had done remarkable things with the soil and water at the pond. All around, inside the pond's sloping bank, the raw earth was carpeted with ice fibers, stiff but crisp and breakable. They rose, like the tufts of a brush, from a soft layer of loose clay. The individual fibers, two inches long, were pressed together like the clinging threads of a freshly sheared sheep pelt. They were heavier, of course, and flecked with reddish-brown clay pushed up with them as the water froze.

When I broke off a handful of ice fiber, the underlying red-clay base came up with it. Under my feet the carpet crushed, dry and powdery. The ice particles sparkled in the cold, bright sunlight, and the earth at the base was beginning to gleam wetly, too. Within a short time the moisture would be released from its icy stiffness and would run down, as far as it could, in tiny rivulets. This is one of the ways the topsoil gets away from the stony, steep sides of unsodded hills.

Around the postholes that Dick had dug earlier to fence the pond, the earth had frozen into little brown doughnuts of fiber-ice and clay. This was the measure of the distance water rises in the ground. In winter, rising, it had frozen. In summer, even in time of drought, it rises in the night and the hard packed ground becomes damp by morning, from the amount of water drawn up out of the hard subsoil.

Loose earth allows water to rise more easily, and the cold night had held it there, freezing its record.

The small patch of ice-clay carpet I carried back with me to the house soon melted in the warmth of the kitchen. Melting, it lost its air of mystery and became merely a thin layer of red-clay mud, covered by a tablespoon of water and easily scraped off onto a potted fern.

The one-horse wheat drill

IN THE first place, buying the one-horse wheat drill was pure hypnotism. We had no work horses.

"Oh, I'll borrow one from Carr or Jim," said Dick blithely. Carr still had old Pat and Mike, the big Belgians that he used once a year in the fall festival parade and once in winter to pull a sled. He has three tractors for farming.

Dick had gone with Warren, the truck driver, into the northern part of the state, where between rows of unpicked standing corn the fall-sown wheat was tall and thick and greenly beautiful. It had been sowed with a one-horse drill.

There was a sale soon after, at which a one-horse drill was advertised. Inevitably, Dick stopped at The Bank on his way to the sale, and borrowed fifty dollars with which to buy the drill.

It was then too late to sow fall wheat, so he stored the treasure in the garage, as a result of which the back end of the car stuck out into the rigors of weather all winter. In spring, at my pleading, he took the drill up to the barn. There, every day for two years, he had to walk around it to get to the ground-feed barrel.

This diminished the hypnotic trance, and he mentioned to a few chosen neighbors that he was willing to sell the drill.

But no buying stampede resulted. A year later he began in earnest to try to sell it. The only person at all interested was Blaine Kerr, whose small farm is a couple of miles farther west on this road, nearer town.

Blaine farms with a couple of thin old horses that are no more enthusiastic about getting the crop in than he is. Blaine is moderately tall, slow-moving, slow-speaking, skeptical, with a peculiar, embarrassed way of clearing his throat, laughing, and ending his sentences with "like" that is not in any way borrowed from beatnik language.

His face looks surprisingly like Herbert Hoover's. He adheres scrupulously to his ethics.

Blaine doesn't own a car. He walks the five miles into town, carrying a two weeks' accumulation of cream in a five-gallon galvanized cream can. That is, he starts walking; usually a neighbor picks him up on the way. The neighbors bring home groceries for him, too. A loaf of bread will just fit into his old-fashioned top-lidded mailbox, in case the neighbor doesn't get home before Blaine has gone to bed.

In a house lighted with kerosene lamps bedtime comes early. It saves fuel, too, in winter.

When he married, late in life, Blaine bought the little four-roomed house near the road because his wife insisted on it. It was probably the greatest amount of money Blaine ever spent at one time in his life. The house has no electricity, no water. The neighbors cleaned and patched his cistern, but for some reason he continues to carry creek water for his wife to do the laundry in, and drinking water from a neighbor's kitchen.

My neighbor, Blaine

Blaine's former house, in which he lived with his mother, is back on the farm, hidden among trees and buck-brush and blackberry briers. He likes its seclusion. It is a two-level, pre-Civil War style, with fireplaces; smokehouse; weaving room; great, beautiful stone chimneys; barns; sheds; summer kitchen—and all the buildings are falling to ruin. His mother's furniture is still in the house, along with books; magazines; stiffened, sweat-stained harness; farm machinery; and the cream separator he still uses. Blaine never went to high school, but reads thirstily. He goes back to the old house to read, so he won't be interrupted by his wife's opinion that he ought to be farming. At an auction he once paid fifty cents for a bushel of obsolete medical books because, as he told Dick: "There's lots of good readin' in them yet–like."

He brings his thin milk cow here to breed to Dick's bull. He always ties her modestly to a locust tree some distance from the house and comes on up to the house to discuss the weather. I doubt that torture could induce him to mention "bull" in the presence of a woman.

For all his eccentricity, Blaine is a man of integrity and acid honesty. He is unwilling to commit himself to any opinion, never parts with any personal possession, guards his personal affairs, especially his financial affairs, with a passionate reticence. Giving a neighbor money with which to buy him a loaf of bread, Blaine takes out his worn, black, pouchlike snap-shut purse, turns his back to the neighbor, and digs out a coin. He has never made enough money to have to pay income tax, but he has to file a return. Dick was the only person he could bear to discuss this with, and even the necessary personal revelations it entails cause Blaine acute suffering.

He wanted the drill and Dick knew he wanted it, but Blaine made several trips here before he even mentioned it. Finally, having cleared his throat, he asked: "What'll you take for that drill–like, Dick?"

He refused the first offer. After several more weeks of negotiating, Dick, who had long since repaid The Bank the fifty dollars he had borrowed, offered the drill for thirteen fifty. Blaine went home.

At the last, Blaine cleared his throat, laughed, and asked: "Could you throw in a little something–like, Dick?" So Dick threw in a worn-out pipe wrench and an old wooden neckyoke, and Blaine said he would take the drill. Dick then spent two hours cleaning and oiling it, and paid Warren two dollars to deliver it to the old pre-Civil War house. It's still there; Blaine never used it either.

LAST WEEK livestock that was out on pasture had to paw away ice to get down to the green bites. The sight makes a farmer cringe. The greening sod is too tender to sustain sharp hoofs. Right now a farmer measures his husbandry not by the number of cattle in the pasture but by the number under a roof, eating hay.

Dick had brought down a basketful of his best hay to show me. "It's just so nice I hate to feed it," he confessed.

Springy as a cushion

It was from a late cutting of the northwest field, had been cured without getting rained on, and so had retained its green brightness and fresh fragrance.

When we came here, that field was mostly gullies and scrub corn.

Dick took out wisps of the various grasses from the basketful and handed them to me with the tenderness of a grandmother displaying snapshots of her grandchildren. "This is the last cutting of orchard grass," he said. "The first cutting gets the heavy stalks, you see, and they don't grow coarse again."

The pliant green blades, wide as raffia, were firm and springy. "Like a cushion," he said. Brown heads of red clover rose above the rest; stalks of alfalfa were discernible. Dick picked out a piece of sour dock. "See," he said, "even the weeds are bright and springy."

Even in drab February a good hayfield never looks drab to its farm owner. He sees its brown realism always overlaid with the green promise of what he's going to do for the field "come spring." Hope springs eternal in a handful of hay.

"LOTS OF THINGS solve themselves if you just let 'em alone," Dick told his son the evening they were trying to figure out how the son could get to a tractor-maintenance meeting in one town and the father to a Fair Board meeting in another town at the same time in the same car.

As usual, he was right. Within ten minutes a neighbor stopped and invited Dick to ride in to the Fair Board meeting with him.

THE VANILLA SALESMAN gets around regularly about once every six weeks. In addition to extracts he sells a complete line of spices, medicines, nose drops, cough syrup, liniment, fly spray, vitamins, pudding mixes, and cosmetics. He has had this route for many years.

In this luxurious era of living, when a farmer thinks nothing of going to town for a bottle of aspirin and the feed truck will deliver livestock supplement or ground feed any time of day, a salesman must have something more than ordinary personality to make a living on a country sales route.

The vanilla salesman has it. He is middle-aged and a little hard of hearing, a middle-sized man with bright, dark eyes, a large family, and a quick laugh. He says he developed his sales policy from watching his wife deal with door-to-door salesmen. When she says, "No, thank you," she means it; and further insistence from the salesman accomplishes nothing except the loss of her good will. So he does not urge me to buy cough tablets or nose drops, when I explain that the family is healthy and we use only the few medicines prescribed by the family doctor.

When the vanilla man arrived yesterday, I was busy at the telephone; but Dick was in the house and invited him to sit by the wood fire in the back room. The vanilla man settled down happily in the maple rocker. By the time I had finished at the telephone, the two men were launched in a comfortable visit. They accepted my offering of Christmas candy and discovered that each, in his time, had been a mail carrier. The experience of carrying the mail seems to create a peculiar sympathy between men.

While they talked, the vanilla man opened his two black satchels and showed me his merchandise. Dick told about his experiences with cross dogs, and the vanilla man responded with stories about queer customers at the money-order window. Eventually they discovered they had mutual acquaintances. Between revelations and discoveries, the vanilla man made out my order for a small bottle of vanilla extract.

The conversation flowed on companionably; the two men agreed that letter-carrying had given them valuable understanding of human nature, but each felt that his present work was the perfect one for him.

Suddenly the vanilla man discovered it was nearly time for the school bus, and he left quickly. A few minutes later the school bus stopped. The children rushed in excitedly. "The vanilla man is off in the ditch. He pulled out to let the bus pass and his car slid into the mud."

Dick got out the tractor. It was the kind of chore one farm neighbor does for another as casually as passing the salt at dinner, but this time it was not a casual gesture. It was a gesture of fraternity. It was in the old tradition; nose drops, like the mail, must go on.

❦ IT WAS GOING to be a dark day, but the sun came up anyway and thereby accomplished something remarkable. The sunlight spread out along the east in long layers of bright-pink crystal light. Its passage was blocked by a heavy threat of rain, but the pinkness simply ducked down and squeezed under the gray like a little pig squeezing under a fence. It came out far beyond the house, on the west clover field, and there lay all over the upsloping field, a pale beautiful pink stain.

❦ FRED COWDEN'S dump truck, pouring crushed limestone into the driveway, goes backward and whines softly. It sounds slightly like the "Pull over to the curb" siren, but much more like the sound of manna falling from heaven. Until you've seen it spilling twenty-one tons of crushed limestone into the muddy driveway from the road to the barn, you cannot realize what a glamorous voice a dump truck has.

❦ THE QUAIL FAMILY, nine in single file, crossed the lower edge of the front yard toward the shelter of elderberry and blackberry brambles. In their quiet mottlement they were hardly distinguishable from the unraked dead leaves they were walking on.

They glided with swift grace into the thicket, like a large family arriving late to church and getting into a pew with the least possible notice.

At night the quail sleep in a deep, scooped-out place in a bank, under a tree along the east road. They are hidden there by a tangle of weeds and wild-grape vines. But there is a red fox's crossing near there, and for extra security the quails sleep in a circle, with their heads pointing out.

The red fox is partial to field mice, which he can overtake by fast pursuit and quick, high leaps. When he wants rabbit or quail to eat, he has to work by stealth, and hopes for a place where protective weeds and briers are scanty.

The Wildlife Federation's book says: "Predation is a natural process, and biologists believe the way to more abundant wildlife is not through the killing of predators but through the management and strengthening of the environment."

ON THE WAY up to the woods today I went past the shallow swag by the walnut trees, and seeing that the abandoned silo door was still there, I couldn't resist a sudden impulse to look under it. Maybe that skink had come back there this winter . . . the one Joe tried to show me three years ago.

Only the Lord, who knows and watches all His creatures at once, would know why a high-school-aged farm boy would just happen to look under a wad of cold winter mud under an abandoned silo door in a shallow, dry sinkhole, and there find a lizard.

When the boy reported this to me, inviting me to come and see the lizard, I was unable to go right then and regretted it afterward. When we went back the next morning, the lizard was not there. Let this be a lesson to me.

Under every door there was a little heap of yellow corn, stored there by a ground squirrel. He had eaten some, leaving the faded, violated hulls. From having been immersed in water several times during the winter, the grains had swollen to twice their normal size and softened, and were probably thereby made more palatable to the squirrel. Very likely he took this into account when he stored the grains there.

Fortunately for the lizard, which is not aquatic, water did not stand there all winter.

"He was the biggest lizard I ever saw," lamented Joe, persistently looking, "as long as my hand, and so black and shiny he was almost blue. Are lizards reptiles?"

"We'll look in the reptile book."

"His head was flat as a pumpkinseed and about that big. Or as big as a tadpole. It was shaped like a tadpole's head. And the end of his tail was as big around as a spike nail. He had two short legs a little way down from his head and further back two even shorter. He was sluggish; he didn't even try to get away. It was all he could do to pull himself back when his back part slid off of my knee. He just barely opened one eye, the left one. I put him back, with the mud over him, where he was when I found him."

It was a great sorrow that we could not find the lizard.

Through that one sluggish, half-opened eye the lizard had probably seen enough to realize that his days of undisturbed hibernation were over.

"His ribs showed pretty plainly," continued the disappointed boy. "He hasn't done much eating this winter."

The ground squirrel probably knew his corn would be safe with the lizard, whose food is insects and now and then a small vertebrate.

We searched in the book of reptiles and amphibians.

"That's him!" exclaimed Joe, putting a forefinger on the picture of a skink. Skinks are lizards; there are twenty species of them in the United States. Active in warm weather, they hibernate under logs, rocks, or abandoned silo doors all winter. They mate in May; the female lays six to eighteen eggs and, unusual for lizards, broods them. The baby skinks hatching six weeks later are about an inch long.

The reptile book says: "Skinks are not easily caught, but will do well in captivity if fed live food, meal worms, ant larvae, or beetle grubs. Keep in a terrarium with rocks under which they can hide." The skink's instinctive wisdom must have warned him; stand not upon the order of your going, but go at once, because this boy can catch anything.

ON A BRIGHT February morning, when the sun throws a warm yellow rinse over the brown earth, the cattle come out and lie down contentedly on the wet ground, or just stand there, motionless, absorbing the warmth through their shaggy coats and asking no further favor of life than this.

The sunlight does them as much good as the feed they would be eating if they were inside the barn.

While I stood admiring them this morning, thinking how perfectly they expressed my own feeling of contentment, a flock of birds of many kinds—wrens, sparrows, snowbirds, and some cardinals—suddenly darted, like one united family, toward the refuge offered by an apple tree near the barn.

Looking up I saw why; sailing above the henhouse was a goshawk, a large, gray-white, graceful bird, scattering fear like a farmer scattering shelled corn on the ground.

By WASHINGTON'S birthday, says the National Wildlife Federation, "the northbound robins are on the move and the great pageant of spring is on the way."

A person walking along a country road can see in bright bark and tight bud the unmistakable statement of intent to participate in that pageant.

By summertime, unfortunately, many of these eager participants will be greeted by a barrage of roadside spray, giving the summer an early-dead appearance and an unpleasant chemical smell. What the effect of roadside spraying may be on insects and birds and small wild creatures that would have dwelt under the vegetation is not pleasant to think about.

Biologists say the danger from herbicides and pesticides is greater than from atomic bombs.

"We do not subdue nature ever," remarks the Federation gravely. "Nature always wins the battles." For this fact man can give solemn thanks.

When the early, northbound robins get here, they will be observed by conservationists fearful of the bad effects of roadside spraying. For the danger is in earthworms that come up temptingly at night, exposing their long, pink and gray, moist deliciousness to the hungry robins. Earthworms store up DDT in their bodies; but robins do not realize this and, unfortunately, there is no way to warn them.

THERE ISN'T much good you can say about starlings, so let that good be said clearly.

In the country they eat insects.

In town they gather around the dome of the county courthouse every evening just about twilight. They are noisy, obtrusive, disrespectful. They pay no taxes. They use up taxes by making it necessary for the courthouse to be cleaned.

When I am shopping in town at evening, however, I like to prolong my errands until the dimming light brings the starlings back to the courthouse. There is a genuinely dramatic eloquence about their return.

They come clutterously, noisily, not like wild geese in poetic formation, led by wise leaders and musically. The starlings come chattering and screeching, in little haphazard groups like crumbs tossed into a strong wind. All the time they keep exclaiming to each other good nights or threats or gossip of the day. Their voices outbid the sound of passing traffic around the square. They circle the courthouse and perch disrespectfully on the gilded metal fish above the dome and on all the stone window ledges aroundabout. Through their clamor one recognizes the sound of pure joy at being back in their safe, familiar place for the night. I know just how they feel; a farmer feels the same way when he gets back to the farm from town.

And then, besides, I don't know of anything that gets more good out of a little color than starlings do, by polka-dotting it over their darkly irridescent feathers.

WE CAME BACK from taking a guest to the bus station, and the house was filled with the particular emptiness a beloved guest leaves even after only a few days at the farm.

This is how it always is: you all go out together to the bus station, but only you come back. And find the eloquent remains of a pleasant visit—the coffee cup by the piano, ash trays unaccustomedly full of cold cigarette stubs, an extra chair still drawn up to the kitchen table, and the borrowed apron thrown over the back of the chair.

Everybody looked around, measuring the emptiness, and then somebody exclaimed: "Oh, I miss Pansy!"

After that, always, the emptiness begins to draw together, like the bark closing in to heal a wound on a tree where a limb has been broken. Every really successful visit is accompanied by this little wound and the grown-together, healed scar afterward.

"WRITE DOWN 'ROOTS,'" said Dick, dictating his list of town errands.

"Oh, are you going to give away those maple roots?"

protested Joe, who would like to keep all these things forever.

"Yes," replied Dick firmly, having already had it out with his own reluctance.

The roots they meant were those of a small maple that had come up in the yard above the crown of the cistern and had been allowed to stay too long, despite my objections.

"It makes the finest place to hang the towel when you wash out at the cistern," said Dick, ignoring the fact that the towel blew away in every summer storm.

Later he said: "It makes the finest place to hang the dipper." Water pumped directly from the cistern seems colder and sweeter than water from the kitchen faucet, out of a storage tank.

Two years ago a storm broke the top out of the sapling, and we cut it. Since then the long roots have been hanging on the back porch, along with ears of the year's best yellow corn tied together by the shucks, and several empty birds' nests and the forked peach limb with which Emerson Dutton, community water witch, traced the vein of water from Benny's old well across the yard down to the spring by the swamp maple.

The sod above the cistern top was thin, and the sapling sprouting there had to go much farther than most trees to get a drink. The length of its roots showed what a tree can do to reach water, and this was why Dick was taking them in to the Soil Conservation office. From the underside of the cistern top the threadlike roots hung down, like a brush, six and a half feet to touch the cistern water. The brush was as big around as a stove pipe. When we cut them, the roots were covered with a dewy vapor, and barely dipped, at the end, into the cistern water.

After two years on the back porch, they were stiff and dry. While Dick finished his morning chores, getting ready to go to town, I hung the dry roots out in the peach tree, hoping they would absorb some of the chilly February rain and be less brittle. They hung there, mournfully swayed by the February wind, like long, scalped tresses of hair.

AT EVERY POINT in dealing with black walnuts you must have some fascinating thought by which to keep your mind away from the fact that anything you get from walnuts you have to work for. Gathering them, in the first place, you can concentrate on the pungent fragrance of their rough green hulls, or the beauty of the autumn hillsides, or the hope that you will hear a flock of wild geese flying south.

A couple of weeks later, carrying the hulled nuts in from the henhouse roof, where they have been washed by rain and dried by sun and wind, you can concentrate on the goodness of walnuts in fudge, cake, cookies, or salted tidbits, raw or toasted. Before you can make any of these delicacies, however, you have to crack the walnuts and pick out the meats.

Get two gallon-sized pans, one for whole walnuts, the other to put them in as you crack them with a hammer on the back step. Some people prefer to crack nuts on a block of firewood with an old-fashioned flatiron, because they can do it in the house. I prefer the hammer and back step because the stone back step has a little hole in it which helps keep the walnut in place while you smash it with the hammer.

A walnut is not truly round, and it must be broken by striking the slightly flattened side, to prevent the hull from separating into its two halves, which are joined, in growing, by an invisible seam. If the walnut halves are struck apart unbroken, the kernels are hidden behind hard inner walls in which there are only small holes, too small for the kernels to be taken out even with a nutpick. (Cracking hickory nuts is exactly different. You strike the narrow edge in order to get out the unbroken halves of kernel.) If you smash the walnut hull properly, the kernel will be exposed in big broken pieces, quite accessible.

As soon as Rose hears the first crash, she comes to the back step and waits in confidence for the bites she will get. She loves the taste of walnuts. It may be that Rose sometimes has the same gnawing strange need a farmer has, when he is hungry for something but doesn't know what it is, knows only that it isn't either candy or fruit or raw vegetables —and it always turns out to be black walnuts.

As each walnut is cracked, I lay it in the other pan, and by the time the first pan is empty, Rose and I have both had many delicious bites of walnut and, by accident, also a few shreds of the black and bitter dry fiber on the rough, deeply grooved shell.

If you get tired of thinking, you can sing; it won't deter your eating. I take both pans into the house and come back with a broom, for the bits of shell left on the step are without mercy. There are regular nutpicks you can use, and we have some, from a nutpick set Carol won as a prize at a 4-H party. I use one of these when I can find it. You can use a ten-penny nail; you could use a coarse crochet hook or even a knitting needle. In the long ago when all farm women had long hair and kept it up with long, heavy hairpins, a hairpin could be used for a nutpick and often was.

From half a gallon of whole walnuts you should get a good jelly-glassful of meats.

AT NOON TODAY Dick brought in an unmistakable clue about his morning's activities. I went ahead putting dinner on the table, but marveling, as always, that on manure-hauling days farmers never suffer any impairment of appetite.

Corn bread and apple butter, boiled beans, onions in vinegar, fried ham, and Dutch apple pie are enjoyed just as heartily on those days as on days when the work has been with freshly mowed sweet-smelling sweet clover.

"In these hectic times," exclaimed Dick, throwing down the newspaper to come to the table, "the only time a farmer really knows what he's doing is when he's hauling manure."

On those days a farmer goes about enclosed in a sense of rightness. He takes his family's living in his hands when he decides whether to sell the steers now or borrow $1,500 more from the bank to finish them out for market, and risk a drop in the market price. Everything a farmer does is gamblous, uncharted. Shall he set fifty acres of tomatoes, risking a hot, dry summer and a shortage of hand labor to pick them? Shall he put his own and his rented creek-bottom acres to soybeans that may be inundated by Beanblossom creek just when the beans are ready to combine? Shall he mow all twenty-four acres of a field and risk getting it rained on before it cures, or shall he cut half and lose the extra time? From every gamble he extricates himself as best he can.

But in hauling manure he can never lose. It is life to his fields. Regardless of whether the crop is surplus and gluts the market and swamps the national farm policy, or scarce and precious, the soil has to be maintained. A manure-hauling farmer knows he is doing the right thing.

The only trouble is, everybody else in the community, and everybody driving past, knows it, too. On windy days, a manure-hauling farmer couldn't attract more attention if he were driving a steam calliope.

MR. OSBORNE, in a nostalgic mood, was talking about whittling. He described it as "a pastime that filled many a leisure hour and sometimes developed into a distinctive art."

He remembered how when he was a boy he knew farmers who always had a ten-cent, single-blade Barlow knife, honed to a razor's edge, and how, after a few whets on the sole of a shoe, the owner could actually split a hair with the knife.

"There were two kinds of whittlers," he said. "One kind did it leisurely, on any old stick or chunk that was handy, just for the sake of seeing the slivers fly." Mr. Osborne chuckled. "I was that kind," he added.

Then there were the artists, who carefully selected just the right piece of wood for a masterpiece. Straight-grained poplar was a favorite, with white pine or cedar running second. From the hands of these artists the little Osborne boy watched marvels come—unbroken chains, carved dogs, horses, little stiff dolls, and the ever fascinating jumpin' jack.

Once one of the artists picked up a piece of cedar shingle and began to slash away at it. When little Gilbert Osborne asked what he was going to make, he paused, looked up keenly, and said, "Nuthin'. I'm just behind with my whittlin', 's all."

Every farming community has its whittler. In this one he is not an artist, that is, he makes no chains or jumping jacks. He doesn't actually care what the stick is: it can be a maple stick picked up in the yard; or a length of clean, shiny cornstalk is excellent, or the edge of your poplar wagon bed. He always has his knife handy, and sharp. On weekdays he uses it for whittling; on Sundays or when he is dressed up, he cleans his fingernails with it. He whittled gently the morning after his father's funeral, explaining: "I didn't want to make a show, nor I didn't want to neglect anything, either." He whittles the day after he has sent his fat steers to the yards and wants the neighbor to ask how much they brought. He whittles carefully when somebody asks an opinion from him about the new minister, or the presidential campaign, or anything else to which he wants to give a noncommittal answer.

He whittles solemnly, thinking up a good joke, and when he has it he looks up casually and inserts it into the conversation like a woman sliding a pan of cookies into a hot oven.

For him whittling is not an art, but an accessory to the art of conversation.

THE LAST DAY of this month is a day of personal inquiry and thoughtfulness for me. It is the anniversary of the day we moved to this farm.

The farm, being in the center of the community, near the church, and having belonged for many years to one family, is a landmark, which gives the neighbors, to some extent, the right of dictation over it.

The soil, perishable Hagerstown silt loam, is hilly and erodable. It is fertile now, but when we came here the farm was a discouraging, run-down place. Now it has the modern blessings of electricity, bottled gas, a bathroom, hot and cold water under pressure, crushed-stone driveways, a new barn and corncrib, house insulation, nonleaking roofs, inlaid linoleum on its uncorrectably unlevel floors, fences, deep fertility, terraced fields. The natural rise in land values, coupled with these improvements, probably makes the farm worth four times as much as we paid for it.

Whatever it is worth per acre, the neighbors are worth more; and this is one of the things I think about on the last day of this month.

Shortly before we moved here from a much larger farm fifty miles north, we received a letter from one of our neighbors-to-be which I have kept ever since because it is a beautiful letter. She was sixty-three years old then. She had gone only as far as the third grade in school. Writing was harder for her than a day's hoeing in the garden, or putting out her snowy, boiled washing, or getting a threshing dinner.

She wrote:

> *Dear Friends, Writing a word of welcom to our neighborhood. A lot of work but hope you will like and prosper. I feel like you will be a help to us and we will try to be to you. Let us know when you are coming and we will go down and build a fire for you. Excuse this misspelled letter. Goodbye and God Bless you,*
>
> *Rena J. Dutton.*

The day we moved was a snowy, sunless day. The rest of the furniture, stock, and equipment had gone; I watched the big old grand piano being carried out and put into a truck, with the snowflakes sifting slowly down into its shining walnut heart. Then I went back into the empty house and sat down on the bare floor, holding our two-year-old boy asleep on my lap. In the room nothing was left of our life there, except the unspent calendar on the wall and my reluctant thoughts.

I hadn't wanted to move. The larger farm was full of happy memories for me. The little boy had been born while we were living there. But the larger farm was too heavily mortgaged, too badly run down. It was the time when nationally the shift was being made from horse-drawn machinery to tractor power. We couldn't make enough money from that farm to pay for it, build it up, and mechanize it simultaneously. But there was an even stronger motive for moving: we had lived in the Maple Grove community once before, on a rented farm, and Dick had never been happy about leaving. The little farm we had just bought was the only one then available, so we were going.

When we arrived here, it was late at night; there was no electric light to turn on. The piano was too long to go in through the front door and turn at the narrow hall, so we had to take a whole window out of the living room and bring bales of straw to slide the piano on. Nine of the neighbors, including Rennie's husband, Emerson, came to help us get our stuff into the house by lantern light. When the truck of horses arrived, one of the young mares had been so discouraged she laid down in the truck and never got up again. I resented it, because I thought she had no right to indulge herself in what all of us would have liked to do.

We have lived here for a long time now. Carol was born here. She has never known any other home, nor any other neighbors, and now I cannot imagine any others, either. I cannot imagine how we would get along without them.

Mankind needs neighbors more than other farm animals do. He needs them more, and needs them longer. And he owes them more.

MARCH

[MARCH]

❧ ❧ ❧ ❧ IN March the nursery catalogues come, like something from a different world where coddling moth and cedar rust do not corrupt nor borers break through and steal the life-juice.

In these glossy pages, plums are green or purple under silvery frost; the grapes, likewise. Apples are far too splendid to be spoiled by the biting-in of human teeth. Currants gleam like red jewels. By merely looking at the pictures, one feels the rosy velvetness of raspberries, smells the delicate early-peach fragrance from the pink cheek.

There are pears in the catalogues also. Bland they are called, but majestic is the word for pears.

One is instinctively gentle about closing the catalogue, lest the peaches and apricots be bruised and the sweet juice start dripping from the page.

These catalogues are hazardous reading for the person who has not enough land or fence for an orchard. He becomes hopelessly lost in the enchantment, and it takes something like the sudden hiss of bursting water pipes to call him back into the world of reality. Even so, if the nursery people would offer small, ready-mix packages of spray, and do-it-yourself instruction kits, probably more farmers would succumb to the temptation of the nursery catalogue.

On land costing $400 to $1,500 an acre, an orchard is an extravagance. Fruit is cheaper to buy than to produce, assuming you can get tree-ripened fruit when you want it.

But, on the family farm particularly, there is a harvest to be had from fruit trees other than fruit. The bloom lasts only a few days each year, but the delight of its memory lasts much longer. A peach tree in a cloud of pink in May, or an old pear tree like a bride in lacy white, an apple tree unforgettable ever after, or a plum tree with white petals falling like fragrant snow adds something to the inheritance offered by a farm childhood. Every farm family needs at least one old, gnarled apple tree, or a bent-over seedling peach tree for children to climb into and think, or to play in. A farm needs a cherry, too, so the farmwife can sharpen her wits by matching them against the birds. The birds will win, of course, because they can give their full time to the contest and because they can tell the wormy cherries from the good ones without having to pick and seed them, with a hiss and a squirt of juice and worm, into a pan.

Maybe the farm orchard ought to start with two cherry trees, one for the birds and one for the farmwife. And after that there would be no stopping.

POOR RICHARD's Modern Maxim: "Next to a native son in the legislature, the best thing a community can have is a good cheese and macaroni baker."

THE CONGRESSIONAL Committee on Agricultural Policy has "concluded that government programs emphasizing production controls will be needed for at least ten years longer." It foresees "little hope of farmers themselves being able to work their way out of present problems of surplus production, unstable prices, and income levels below nonfarm levels, without government assistance."

While the Committee was coming to these conclusions, the USDA was also busy. Among its findings and statistics, the Department of Agriculture discovered that last year nearly three million farm residents, more than a third of them women, were working at nonfarm work. The USDA estimated that out of every three dollars earned by farm residents one dollar comes from off-the-farm jobs.

Part of this income goes into taxes. Government assistance is paid for out of taxes. So it appears that farmers, by working off the farm, are able to help the government finance its assistance to them. This policy might be called robbing Peter to pay Peter, and is not the same as robbing Peter to pay Paul.

The system of living on a farm and working away from

it, whether to pay taxes or to meet general expenses, is one of the ways farmers have worked out, for themselves, to help themselves. The reason they don't "go the whole hog" and get all their dollars by off-the-farm jobs is that they are farmers at heart, just as an oak is an oak all the way through, and the pull of the land on them is too strong. Since the earliest days of farming, farmers have had to make some sacrifices to remain farmers. Probably they always will have to.

The American Farm Bureau did not agree with the Congressional Committe's concludings, and suggested a different approach which would "set supports that would discourage surplus production."

All of this conforms to statements made by run-of-the-plow farmers that as long as the government guarantees a high corn support, farmers will decide at corn-planting time that the government is their buyer and will plant accordingly.

Hidden somewhere in the clutter of all this finding and concluding and free-spieling is the old-fashioned belief that it would not ever be really good for any generation or any group to have all of its problems solved for it by the government, even if the solutions really worked.

AT DUSK last evening, Joe discovered a little gray screech owl high up on a stack of hay bales in the tool shed. He leaned a long ladder against the hay and went up quietly. When he grasped the short, feather-covered legs gently, the owl was not frightened. Neither was the bird frightened when it was being displayed to us at the house.

Like a traveler in a foreign land, the owl took interested note of his surroundings, veiling his curiosity under an air of kindly politeness. His eyes, round and yellow, are actually not much larger than a dime, but looked large as dollars because of their extreme clarity. By opening and closing his hooked gray beak with a sharp, cracking sound, the owl made his self-sufficiency known.

When he opened his mouth, we saw that its lining was a watermelon pink. His tongue, the same color, made a tapering, sharply-pointed trough. Sitting alert and upright, he was eight inches tall. He had two little pointed feathery ears. When these were gently pushed forward by a curious hand, two little apertures were disclosed in the feather-covered head. The screech owl has sharp hearing.

As I stroked the back of his head, as one might stroke a kitten, I thought he made a barely audible purring like a kitten that has lost its voice and purrs in a tiny whisper.

Under the depth of gray feather, the hard core of owl body was surprisingly, appealingly, small.

Released in the room, he flew about, perching sometimes on a mirror's edge, sometimes on a tall cabinet. We hoped he would utter his distinctive screech, which is a plaintive whinny, or a quavering vox tremolo, but he made no outcry at all.

Joe put him finally in the bathroom, where there would be fewer old dishes and lamps to knock over. We wanted to keep him in the house until Dick and Carol came home to see him. When the bathroom door was opened, later, the owl came out as if he understood his assignment, and readily let himself be caught again and admired. By the time the boy returned him to the tool shed, it was dark enough for the owl to begin his earnest hunt for supper.

"I wish I could have kept him and tamed him for a pet," said the boy wistfully, whereupon I suddenly realized that as yet this year the farm has not acquired any of the unusual pets which enliven our summers. Joe has given loving attention to many: a turtle; a baby raccoon who never would accept charity; an orphan ground hog rescued from solitary starvation; an amiable snake; a mole taken away from Rose; a lame pigeon; a mouse in a glass fruit jar; seven frogs in a glass jar and a single one as long as a loaf of garlic bread; a nest of baby rabbits; two featherless, pink sparrows fallen out of a nest; a pair of somnolent snails; two delightful borrowed goats with a miniature, faithfully copied oldtime farm wagon and harness; a flying squirrel named Maynard; a beautiful skunk.

All this is part of farm living, by which man acquires a sense of his place in the world, and from which he gains something of pride and also enough of humility to make his kind bearable.

LATE SNOW lies in little patches against the hillsides, like scraps of dough left over from cutting out cookies. If it were all squeezed into a ball and rolled out again, there would hardly be enough to fill one more baking sheet.

The beginnings of violets

There were two bluebirds on the west fence this morning. Frogs sang last evening from the swampy corners of the fields; and under the maples along the drive, where violets overtake the yard every spring in an overstatement of purple, the beginnings of these already show. As yet the beginnings are only the horny roots above the ground, thick as a child's middle finger, but the rolled-up tiny red splinters on them are unmistakably the incipient heart-shaped leaves of violets. I put one in a glass of water in the kitchen last week to make sure.

Today, when I went down to the mailbox, there was a fly sitting on top of it. His long wings were folded against his sides, and he seemed uncertain what to do. When I pulled down the mailbox door, his problem was settled: he slid off into the cold oblivion of tire tracks in the mud.

It is near the end of winter, but not yet spring. We know this because, traditionally, between the end of winter and the first of spring, school children "come down" with something. They take it like corn popping—first one, then three, then several. Suddenly the list is long; mothers quit hoping their children won't take it and just hope they'll get it over with quickly.

MY HOUSE and I are, as Dick says of himself and his milk cow, "the very best of friends." We understand each other. Neither infringes on the other's rights.

It was not always like this.

For a while a feud existed between us. My house had its

conception of the ideal farmwife, and I was not it. I had my conception of the ideal farmhouse, too. My ideal looks old and used, and contented—a gracious, hospitable, vigorous place. There is a big barn showing through trees behind it. The farm has belonged to the same family for many generations. The children have lived there, grown up there, gone away to school, and have been sustained ever after by the memory of a rich, happy childhood spent on the farm. My ideal farmhouse has a petted look; its fur is smooth and glossy from much good food and loving brushing.

That was not this house when we moved here. This was a resentful, vindictive old house, not very well built to begin with. It entered upon old age wronged, embittered, suspicious, demanding to be let alone.

Such an attitude is as unnatural in a farmhouse as in a farmer, and can be brought about only by the same thing, unhappy experiences with women.

The house started life happily as a one-roomed schoolhouse, the Wampler school, up on the hill on this farm, in the woods across from the Maple Grove church, with a road running in front of it. It was bought by a man who had, in his long, happy-go-lucky, spirited lifetime, five wives.

When he married the first wife, his parents were already living in an old brick house down here in the yard, which has long since been taken down. He bought the old Wampler schoolhouse, moved it down the hill because the water was here, and made it into a house. Each woman, in her turn, attempted to remodel the house nearer to her heart's desire. The result was a superb architectural blunder that could not possibly be corrected.

The house brooded, sulked, and remembered its happy school days.

When we came, there were eight rooms, two of which were kitchens; a pantry that had three walls taken up by doors; two halls; three porches; an unfloored, windowless attic; a basement under one kitchen and porch. No heat, no plumbing, no electricity. Two telephones; one maintained by farmers, and giving only neighborhood service, the other giving service from the city exchange. A delightful fireplace, quick to take hold in flame.

The room we wanted to use as a kitchen was entirely surrounded by other rooms. Its one window peered nearsightedly into another shedlike room that had three windows, two doors, two cabinets, and hardly enough width to walk past a bed. We tore away this room so that we could find our way through the kitchen in daytime without carrying a kerosene lamp.

Entering the house by the front door, we came in through a hallway too narrow for any furnishings except claustrophobia. We picked our way through a dark room to a pantry, which was wedged between two kitchens, and went on to an attached woodshed, from which the previous owner had sawed out one wall to make a garage into which no car could make the turn.

Looking for a restful place to sit down in, we usually wound up in the barnlot, building fence. We quit trying to come in through the front door. In most farmhouses the front door is, at best, only a vestigial structure like the human appendix, anyway.

Entering the back way, we came first to a sharp-edged,

very white stone step beside a long, trembling porch. From the step we cautiously set foot on the porch floor, avoiding the cellar door therein, which had a too-cordial way of urging us to drop in. From that porch we went through another, screened, porch to the pantry and kitchens.

Porch remodeling began when a neighbor leaned against a porch pillar, which promptly fell out on the ground. We tore away the outside porch. The white stone step had to be moved, and when turned over, proved to be a headstone carved with the name and birth date of the fourth wife. We set it respectfully aside until we could decide what to do with it.

Turning away from this discovery, I rejoiced to see Eaglie walking up the driveway. Cheerful and chatty, she has lived in the community for more than fifty years. She loves it, knows its history and all the people, and loves the people. She is tactful and loving, but can speak tartly, too, if need be.

She gave me a lively account of how this house was moved down the hill and told the story of the farm owner and his five wives, and where everybody had died. One death occured in the dark room, "just about where your cabinet is, Rachel." One in the back kitchen, next to the woodshed. "She was took bad, and sent for Clint and me," she said. One death occured in the living room—"It was a bedroom then" —and one in the dining room where we were sitting, "just about where your chair is, Rachel." I got up and poked the fireplace and sat down in a different chair.

When it was time for Eaglie to go home and feed her chickens, I walked with her down to the road. Thanks to her visit, when I came back I was feeling gay and energetic. I got the hammer and drove a nail into the wall to hang up a picture.

The angry old house drew back and struck me in the face with a heavy chunk of plaster.

Angry in turn, as I was, I had to stop to soothe my cheek, and stood looking out the west window, up the lovely, sloping field, toward the hilltop, where the schoolhouse had been. Anger gave way to genuine sympathy.

Poor old house! In the short time we had lived here, I had been up on the top of that lovely hill, had seen the wide, free-running view, the green barley fields in spring, the snowy blanketed winter wideness, the dull browns and rich reds and the shaded greens marking off the seasons. I, too, have longed to escape from the foot of the hill and go up to the top, where I could see and smell and even taste the freedom.

And so I thought I could make peace with the house. "We'll be good friends, in time," I prophesied.

It has come true. I love this old house. We are the very best of friends.

❧ CARR STANGER's car came struggling slowly along his stony, narrow driveway. At the road it speeded up with the sound of an old bucket being thrown on the ground, and presently Carr was at the back door, asking for Dick. I could tell by the happy look on his face that he had an emergency of the kind farmers like.

Not the painful kind, as when a pin shears off the hay

baler in a tense race with rain. Just a cozy, hand-sized emergency that requires a car and a neighbor's advice and company for the day. It must come on a pleasant day, at a convenient time, such as when the farmer has just finished sowing his oats and doesn't quite need to start plowing for corn.

That morning Carr had discovered that in the field he tiled last year water was pouring into the creek from under the galvanized tile outlet, instead of out of it. That meant that, someplace back in the field, water was escaping from the tile line and would eventually make a hole in the field. It wasn't going to happen within the hour, nor even within the summer, probably. It was just a nice, obliging emergency that gave Carr an excuse to ask Dick to spend a pleasant day with him driving around hunting a back-hoe.

A back-hoe is a short-stroke hydraulic digger that attaches to a tractor and digs a ditch by which the broken tile could be located. The tool isn't needed often enough to justify every farmer's owning one. It's a custom-work tool, like a bulldozer or a power posthole-digger, and there aren't many in this community. It would take probably a whole day of driving around, on a fine day, to locate one.

Emergencies of this kind help to keep a farm operating smoothly.

MARCH WINDS attacked the heaped-up leaves around bushes and fence corners like a vacuum cleaner suddenly gone mad.

Watching the upheaval as he put on his coat, Dick

smiled. "It's nature's way of uncovering the things that needed protection through the winter and don't need it any longer," he said.

ANYTHING ELSE when you can get around to it, preferably in the right sign of the moon, but sweet peas on St. Patrick's Day and Irish potatoes on Good Friday, whether or no.

On the morning of St. Pat's the hard-frozen ground was blanketed with snow, but it was an unworthy, lightweight blanket, not even hand-washable, and it soon wore out under the brisk sun. Thanks to the Washington Consolidated fifth grade and Kay Morgan, who are selling seeds this spring, I had a packet of sweet-pea seeds on hand. By midafternoon it was possible to dig a trench in the yard, in front of the diminishing woodpile, and plant the peas. I gave them a shot of commercial fertilizer and a pep talk; and now if the Bantams, the friendly but ubiquitous red Duroc hogs, and the cattle will just walk someplace else, and not smell the freshly turned earth, there will be delicately scented, many-colored sweet peas to pick every day this summer.

"That's why I don't plant 'em any more," said my neighbor Edith Fyffe, for whom everything grows well. "You have to pick 'em every day."

THIS MORNING will go down in family history as "the morning we saw more robins in the front yard than we could count."

There were robins in the yard near the porch, and in

the east yard where the boys play baseball, and all over the front yard down the road. I quickly counted up to thirty-two. "But you haven't counted those west of the maple," said Dick. He stopped at forty-eight, but the robins kept on coming.

They were snatching worms from under the leaves. Some pushed the leaves impatiently aside; others picked them up and set them aside. They left little dark leaf mounds, as if an electric mole had started to tunnel there and his current had gone off suddenly.

"A bird eats his weight in worms every day," murmured Joe.

The male robins' red was a hearty, brownish tone; their caps were black; they seemed to be wearing all new clothes. They had a brisk, impatient manner, not arrogant, more like the self-sufficient manner of a good truck driver. The females were much paler, but by no means insipid. They looked as if they had dressed carefully, putting on a fresh white blouse, and over that a salmon-colored shawl that parted at the center, disclosing the blouse all the way down to the long, gray tail feathers. Their soft gray upper feathers were like gray jackets, and their heads were capped with the same soft-textured gray. Even when the wind blew their feathers backward, the robins looked tidy.

While they scratched, they kept up a song. It was a short, sharp single syllable, almost like a bark, and they did not all speak at once. Speaking by groups, they kept the sound continuous so that eventually it was like a ball of sound rolling across a floor. For the most part the birds were congenial, although two young males held a brief argument.

Within half an hour they were all gone. They left only the little, dark, pushed-up mounds as proof that this remarkable sight had really been there.

"It is always fitting," John Burroughs once told a visiting friend, "to preach the gospel of beauty in the commonplace."

Sometimes it takes a child's unhampered imagination to see this gospel.

I had cracked a pan of black walnuts and thrown them out under the althea bush for the birds. Carol, then in the fifth grade, had helped me, and we admired the beauty of the shell's inside profile. If a black walnut is cracked so that one half retains its outline, unbroken, it makes a heart shape, enclosing the polished inner wood. It is deeply indented at the top, with a graceful, flowing curve, and comes to a sharp point at the lower end.

Later she came to me with her palms held shut. "I want to show you a family I have," she said and, opening her hands, disclosed some halves of walnut shells.

"They're all monkeys," she said. "This is the father monkey." She handed me the largest shell, which did indeed bear a startling resemblance to a brown face surrounded by shaggy hair. The deep, black, irregular convolutions at the top of the shell suggested a dark, hairy head. Under the low forehead the two deep, empty kernel channels gave the bright-eyed, vacant look of a monkey's face.

The next shell, being smaller, was of course the mother. As in the standard American family, there were also a boy and a girl. One shell remained in the child's hand. It was half of a dry, immature walnut, lighter of weight, paler in color. Its underdeveloped outside ridges were shallower and less intricate, less monkeylike. In fact, it looked more like a human face than like a monkey's. "And this is a salesman," continued Carol matter-of-factly. "You see, he is blond, and you have to turn the shell upside down to see his face." Only a person who understands the beauty and commonplace of a child's imagination could explain why the man had to be a salesman.

"THERE'S some kind of queer bird down at the spring crying," said Joe in the mild, moist dusk last evening. We stood on the back steps to listen.

They were night hawks. Farmers call them bullbats and recognize their cry as a voice of spring. It sounds something like the cry of a kildeer and, oddly, of spring peepers, too. It is a two-syllable, plaintive call with a note of urgency in it, so that if you didn't know the bullbats were flying of their own free choice, you might think you ought to go down to see what's troubling them.

At night and at daybreak they fly, with their slightly bearded mouths open so that insects will fly into them. They are a mottled brown, gray and white, with a white band under the chin and white on the undersides of their long wings. Bullbats, members of the goatsucker family, are first cousins of the whippoorwill, whose nostalgic cry, coming out of the chiffon-thin dusk, is the very confirmation of summer.

THE EARTH, giving up its winter rigidity, exudes a personal fragrance that is barely perceptible, but exciting. It is not yet the fragrance of blossom or new unfolding leaf, for these are yet to come. It derives partly from the stirring of infinite numbers of tiny lives, both plant and animal, within the earth and close to its surface. The fragrance of this stirring blends with the fragrance of clean-washed air, into which spring lightning has released nitrogen. Something may be added to it from the delicate breath of new grass, thrusting up its first pointed swords, or the little mounds of earth in soap-powder-sized crumbs where worms have drilled down into the ground and out again. Some of the scent comes, certainly, from the warm, moist fiber of winter's old pelt, the dead leaves, dried grass, broken bark, and dead stalks lying around getting wet by rain and drying out again.

The bountiful, abundant breast of earth moves, breathes, perspires; and its fragrance is one of the subtle, exciting fundamentals accompanying the return of spring.

"SPRING COMES," exclaimed Grace Lundy, teacher of third grade in a consolidated country school, "and it's the same pattern, repeated diminutively. Women clean house, and school children want to change seats!"

Farmers grow restless, reform-minded. On the way to

the mailbox on a balmy March morning, a farmwife stoops to pick up a maple limb blown down in last night's high, singing wind. The next thing she knows, it's time to start supper; and she's had a marvelous day planting asparagus, rhubarb, and horseradish, reroofing the brooder house, or fencing the garden.

The passion for reforming overtakes all the people and creatures on the farm. The high school boy hurries home to pull out fence posts from the soft ground with the tractor. The pre-high school girl takes a garden trowel and cleans out the sides of the little creek in the yard, which will be bone dry by May. Old Puss appears on the back walk, proud and hungry, after a day's absence; her underside is a five-signature petition asking for an increase in food allowance. Cows bring forth new calves. The dog brings forth a long-buried bone to gnaw in the delicious leisure of the March morning. The farmer makes a swinging gate, bringing the total of swinging gates to one. The whole family goes into a delicious enchantment, from which they will emerge only at dusk when the air cools and reason gathers, like dew, on the forehead and the radio predicts snow and colder for the next day.

IT'S SOME TIME yet to the sun-warmed bliss of garden-making, but fortunately there is some planting that can be done right now. A bulb to be set out is a delight, here on the outermost still-dormant rim of spring.

The person going out to plant wears gloves on her

hands. She lifts out the first shovelful of earth and empties it to one side. It is neither wet nor hard, merely moist, easily poured. It contains air and well-made air channels and is therefore light in the way good bread sponge is light. Its fresh, earthy fragrance rises up, familiar and provocative; and suddenly the planter knows why she wore gloves. It was for the pleasure of shucking them off and taking a handful of the fresh earth up in her bare hands, to smell its satisfying perfume and feel it against her skin. Now is the moment for an intimate renewal of kinship with the earth.

The handful looks unoccupied, but she knows that if she took it into the house, watered and warmed it, the surface would quickly be occupied by small eager sprouts. She knows that to impersonal nature those weeds and grasses are as precious as her most expensive lily bulb. Replacing the dug soil over the top of the bulb, she thinks how bare it looks. It is difficult to realize how tenaciously every inch of this blank, quiescent earth will presently be fought for. It seems inconsistent that gardening is so comforting a labor.

FANNY and Emmett Dunning live in the Crossroads house. From late March until mid-November, when I pass there I drive slowly, the better to see everything. Fanny's vegetable garden in the stone-walled enclosure next to the road is as beautiful as the seed catalogues. Seeing the tidy rows of cabbage, beans, tomatoes, and things unidentifiable from the road; the square of crisp, pale green lettuce; the rows of zinnias, marigold, dahlias, gladiolas, and plushy red coxcomb, I always resolve to ask Fanny to let me help her make garden, so I can make my own the same way.

Fanny works hard in her garden. She comes out early in the morning, and she uses a hoe. "I never could do much with a garden plow, fur's that's concerned," she says, laughing her hearty, comforting laugh. She thinks the plow is allright for laying off rows, though; it gets them straight. "I just look at something at the other end and start walkin'," she says.

In Fanny's yard there are so many flowers that the little square white house is like something set down in the midst of a party-table bouquet.

Any day now, John might come and plow Fanny's garden, and I want to be sure of being invited to help her. So yesterday Carol and I took a couple of begonia bulbs and stopped there. Fanny waited in the doorway, holding the door open for us, having heard the car and looked out through her crisp lace curtains.

The row of jonquils in her border already have tall green tops. It wouldn't have surprised me to find her roses blooming.

There were enough rocking chairs, with cushions in them, for all of us around the big fuel-oil heating stove in the living room; but Emmett sat on a couch by the window. He rested his hand on his cane, smiled often and laughed, but said little.

There were flowers in the linoleum rugs on the floors, flowers in the wallpaper, flowers in the big cotton apron Fanny wore over her dress. She wears her hair drawn straight back from her face and rolled up in a big white knot at the back of her head. She and Emmett spent the first fifty years of their married life on one farm, then moved here because Emmett was not well and their son John wanted them near.

I could see into the immaculate kitchen, see the old-fashioned kitchen cabinet with which Fanny started housekeeping. I could hear, from the kitchen, the sound of the big modern freezer, purring like a well-fed cat. On the clock shelf, which had a blue ruffle tacked around it, was a black clock with a small round face, short pillars, and, on top, a prancing black horse.

"It's only the second clock we've had in all our married life," said Fanny with pleasure. "Uncle Steve gave us the first one for a wedding present. When it wore out, we bought one as near like it as we could find in the catalogue, and set the horse on it."

"How long have you been married, Fanny?"

"Fifty-eight years in December," she said, but couldn't remember the day of December. She had to go look it up in the big Bible in the bedroom. We both laughed about that.

"I know it was an awful muddy night, fur's that's concerned," said Fanny. "The horses like to pulled the 'shavvs' out of the buggy getting out to the farm."

She said she'd be glad to let me help make garden.

She saves her own seed beans, a cream-colored bunch variety of which she got the start from her mother, who also saved her seed beans. Fanny keeps weevils out of them by sprinkling them with lime; or, if she has no lime, a few drops of kerosene will do. She plants her cantaloupes alongside her cucumbers; they won't mix. In her old garden, back on the other farm, she had rhubarb, gooseberries, and sage against the fence, and she misses them here.

She gave me a start of her white cucumber seed. "Mix a little sulphur in the hill when you plant the seeds," she said, "and bugs won't eat the plants."

ALMOST ANYTHING becomes bearable at the last moments, if you feel reasonably sure they are the last. It's a kind of "OK, but don't let it happen again" forgiveness that makes these last March snows seem pleasant.

This morning the snow falls quickly, fine and dense like white mist. Above the neighbor's sloping field across the road, the sky is pale gray as a catbird's feathers. The gray sky reaches down to the tip of the snow-whitened hill, and there the two are carelessly basted together in long, irregular, black stitches made from the brush-grown fenceline.

Maples reach far up into the sky, unobstructed and bare of outline because the misty snow has obliterated the small details of the familar landscape. On all the dark, shaken, wet branches are distinct little cinnamon-colored bowknots. They are the maple buds, swelling and reddening with the same knowledge that causes sap to flow under the wet maple bark. Sometimes the buds look like delicious ripe berries, and a hungry person can imagine making them into deep cobblers, with spicy steam coming up through their slits, to serve with thick country cream on a chilly March evening, for supper.

"COME OUT, come out! The rain's stopped and everything's in Technicolor!" cried Dick at the time of sunset Monday.

The fields were the bright, unreal green of the post cards you buy on vacation trips to send home.

It was a spectacular view of a familiar occurrence, a farm sunset. In the west, at the top of the hill, the sun had turned the sky into a bowl of melted silver. Trees made long, darkly brilliant shadows. The hill itself cast a long shadow over the yard and field, and beyond the shadow everything was colored to exaggeration.

The real wonder lay in the east. A whole rainbow, in all its subtly varying tones, was hung in the sky. One end rose out of Jim Scherschel's fencerow a mile away; the other end stood firmly distinct on the top of Clyde Naylor's barn, three miles in the other direction. An immense rainbow. At each end of it, the sprout of a secondary rainbow burst from the ground. The sunset light, extending far beyond the fields, touched a meadow on the next farm with the hard brilliance of a hay field under a burning summer sun. The same light touched the tip of a weathered gray barn on a still more distant farm, turning it chalk white.

The long shadow of the west hill climbed up the swamp

maple in the springlot, but did not quite reach the top. At the very tip, the maple's red satin buds were intensely brightened.

The brown of newly plowed fields was intensified also. Torn strips of smoke-brown clouds floated above, across, and under the rainbow.

Then across all this dramatic color, suddenly, three small gray swallows came, flying in a bent line from the south. When they flew into the hard, melted silver sunlight, they were suddenly turned to a startling white. It was almost too beautiful to bear.

A PERSON walking in the woods in these late days of March, or along the leaf-strewn edges of road and field, may discover cocoons, marvels of insulation into which last fall certain furry worms confided their lives and destinies for the winter.

The cocoons are gray-brown like winter leaves, nature having given the worm that bit of insight as part its ration of talent. They lie strewn carelessly on the ground, or hang from trees and weedstalks like old dried leaves that haven't quite got around to blowing loose. They are almost hidden, but not quite, for whenever nature makes provision for anything to hide, she also supplies a few eyes capable of finding it, but not inevitably certain to.

Nature is impartial, and it's hard to tell whose side she's on.

Cocoons

Some of the cocoons picked up by delighted human finders will prove to have been discovered already. There may be little punctures in the tough drab paper, or the ends may have been torn open by moth enemies, not from enmity, but from hunger. The difference between enmity and appetite is a matter of who is eating whom.

In a honey-locust sapling we found two small, pale brown cocoons, hanging round as a bubble-gum triumph well underway between the lips of a seven-year-old. From one bubble, already broken into, hung a long swatch of delicate silken threads, blond and bright and generously sprinkled with what looked like finely sifted corn meal.

In the small hedge tree along the east road we found the egg case of a praying mantis. The eggs are contained in a wad of substance that is crisp as cornflakes and appears to have been made by rolling up narrow strips of muddy creek foam.

We also found the gray-brown paper cocoon of a giant Cecropia moth. It was attached to a dead goldenrod stalk. To my sorrow, it had been punctured on one side; one end had been made ragged like the furry tear on a heavy brown-paper sack. Inside we could see the mass of strong, fine threads by which the pupating life had been attached to its superbly insulated waiting room. The cocoon has weight and when shaken it rattles, so we have hopes of its emergence. To watch a cocoon break open and the moth emerge and slowly unfold its glorious wings is an experience that takes one's mind completely away from personal affairs and is remembered afterward with something of the awe one might feel in seeing a world created.

THE GREENING RAIN came in the night Tuesday. There were some tossed-up coals of lightning; there was a wind that began in a careful whisper as if trying to waken one person without waking a small sleeper next to him. But the wind's joy increased as it sang until finally it was roaring and shouting and not caring who heard.

And then the rain began, the tonic, greening, transforming rain that comes only once a year.

At gray daylight, silver drops still clung to the undersides of peach limbs; but the rain had stopped, and now every living blade and stalk whose destiny it is to be green was suddenly astonishingly green.

This miracle happens every year at the beginning of spring.

This one rain is as distinct and set apart from all other rains as one handwriting is from another. Later on the grass will be thicker, taller, tougher, but right now every blade is as greenly green as it can ever be.

Cattle notice the difference immediately. Farmers do, too. This morning every farm neighbor you meet will exclaim happily, "My, ain't it greened up nice since the rain!"

APRIL

[APRIL]

DICK came into the kitchen, and I could tell from the blissful look on his face that he was about to begin something he had been looking forward to doing.

"Do you want to help me mix grass seed?" he inquired. He had just come down from the field north of the woods, where the fall-sown wheat is now bright green and as tall as the thickness of a farmer's hand, and the ground is cracked open just right for seeding with mixed grass for a new pasture.

We used to broadcast the seed out of a hand seeder which had a canvas bag and a crank that you turned by hand as you walked across the field. If you sow grass that way, you sow it on top of the last good snow, in February. But the broadcast way was never as successful as drilling the seed in from a wheat drill in April, when the ground is dry enough not to roll up into balls. The extra fertilizer that goes on at that time helps the wheat, too; and all this adds to a farmer's pleasure in seeding a new pasture.

From the back of the car Dick had already set out the seven new sacks of seed. Beside them, in the driveway, he had set the galvanized washtub, the scales, a coffee can for measuring. I carried these into the kitchen while he brought the seed. We spread a sheet on the floor.

He untied the small sacks, turning the tops back as tenderly as a mother folds the children's coat collars when the children are in the coats. As he opened each sack, he invited me to look into it.

Mixing grass seed is a job that would delight the makers of textiles. It delights a farmer, too; a great sense of security and richness pours out with the seed.

I dipped up a handful and let it spill slowly back into the infinity of the heap on the sheet. "I hope you're putting on enough," I suggested doubtfully.

"I'm putting a hundred and three pounds on nine acres," he replied, "and you'd think that was enough if you had to buy the seed. There's $60 worth in those seven little brown-paper sacks."

He has kept accurate record of grass mixtures used on each field and therefore knows what proportion does best in each. He said: "It's my ambition to have the water clear enough to drink when it runs off this farm. A two-inch rain will run off muddy, and muddy water will carry away three tons of topsoil to the acre. That's as much as a farmer can build up in a year's good management."

He emptied the sack of timothy onto the sheet. This is a small, lightweight seed without gloss. It is suggestive of brown and white calico with small designs on it. "It helps make sod," said Dick. "It used to be $2 a bushel; now it's $11.50."

He poured out another sack. "This is redtop." I took up a handful of the gray and brown seed, no bigger than a splinter. It had a textilelike sheen and ran out silkily between my fingers. Redtop is expensive now because farmers who used to raise it have discovered the greater profit in raising soybeans and nobody wants to fool with redtop any more.

Another small sack held two pounds of Ladino ($5.50 worth), a glossy, polished-looking golden-yellow and brown seed. A newcomer, it is popular with farmers who have tried it.

Some of the "little red" clover seeds were greenish-yellow; others were blackish purple. These seeds, though of two different colors, would produce identical plants. "Grandpa used to say a bushel of this was the best investment a farmer could make," said Dick. "I don't know what a bushel cost grandpa. It costs $36 now."

There was Korean lespedeza, with a guttural, foreign look. The humped-up, slightly flattened, harsh-looking gray hulls contained seed the color of ripe eggplant. It is threshed just ahead of the first frost. If left on the stalk in the field, it breaks from the hull in February and plants itself, in the efficient, thrifty way of nature.

I dipped up a handful of alfalfa seed, heavy, oily-looking, and let it run out of my hand in an opulent green and gold stream (at $42 a bushel).

The kidney-shaped seed of sweet clover is more glossy, golden-green, and also fragrant, like the hay it produces. A cutting of hay that includes a liberal amount of sweet clover, with white or yellow blossoms, will scent the air for a long distance. A pillow filled with fresh sweet-clover hay will retain its fragrance well past the next haying season.

While Dick went out to the tool shed, I continued mixing the grass seed, plunging my arms in up to the elbow, enjoying the cool silkiness and richness of it.

When Dick came back to the kitchen, he was carrying a limp burlap sack with seed left over from last year's sowing. He removed the binder twine tie and emptied the sack into the heap on the sheet. "Orchard grass," he said respectfully, "the last green bite in the fall, the first green bite in the spring." He shook the sack vigorously to shake loose the chaffy gray splinterlike seed clinging tenaciously to the burlap. It was uninviting to the touch, but I respected it also, because I have seen what it can do. In Carr's back clover pasture last fall there was a streak of bright green against the dead brown stubble. It looked as if somebody had swished a bright-green-soaked cloth across a brown table top. The brown was from dead ragweed that had come up, ripened, and died after the clover was mowed. The green was orchard grass, and it had choked out the ragweed where it grew.

"A pound and a half to the acre," Dick said. "Not more than two pounds. Cows'll eat that tender stuff and leave the coarser stalks of clover and alfalfa. A mixed grass actually makes more to the acre than any one of 'em alone."

We scooped up the mixed seed into the tub and carried it out. I was thinking how much life was contained there; how could the world ever be utterly destoyed with so much grass seed around?

THE CATTLE and the ponies are fenced out of the small hilltop woods almost in the center of the farm.

There is a particular blessing to be had from walking in a woods or a field from which no other harvest is exacted except that blessing. As the population of the world steadily increases and land is nibbled away for public uses, the human hunger for mere space becomes continually more insistent.

Yesterday, because the day seemed right for mushrooms—warm, moist, with gently running wind—I went up to the woods to inquire. But the earth is not yet ready. It needs to be warmed through, then rinsed off with two or three mild rains. Some farmers say it takes thunder to bring up mushrooms.

There should be May apples in bloom, and jack-in-the-pulpit almost ready to start the doxology. There should be Dutchman's breeches, with little yellow patches, and squirrel corn slightly past its frosty prime, and at least three colors of violets. There should be a drowsy snake, well out of the way somewhere near the blackberry patch.

Yesterday had none of these. There were toothworts, whose charm is not in their color, pinkly insipid and made even more insipid by streaks of blue, but in the mere fact that they are there so early. Around the trunks of leafless trees were thick siftings of salt-and-pepper. Their white blooms are so tiny you have to look through a magnifying lens to discover the exquisite design and rich red lining they have. There were several dark blue violets, but as yet no white or yellow ones, no Indian turnip, no May apple, no sweet fern. There was white hepatica in patches the size of a scatter rug. Sunny areas of the woods were thickset with long, narrow leaves of trout lily, and from some of the larger plants the opened lilies made a brightness above the mottled green and brown leaves.

I picked up, near there, a fragment of bird's-eggshell almost the same mottled color. It had been a large egg. The yellow stain of yolk against the fragile, porcelain white inside showed that it had not been broken by a bird hatching from it. It had been stolen, carried away, and eaten.

Wind combed softly through the dark tangle of leafless treetops. It had been a hard winter for trees. Fallen limbs almost blocked the path to the mushroom place. But the feeling of spring, about to be blown gaily in, was all over the woods.

Quiet down there, you frogs

THE CHANT of frogs, from ponds and side ditches and water holes, is the first really official announcement of spring. It has another effect, too. It inspires farm boys to scoop up frogs' eggs and bring them to the house, in a glass jar, to see what happens.

In the kitchen this morning is a half-gallon glass jug from which Joe rinsed out the traces of cider and into which he put five handfuls of frogs' eggs.

They make no noise or odor. A fellow occupant of the kitchen would be unaware of them, except that, having once looked, one cannot quite forget them. In their cider-jug laboratory, they offer a narrow, disquieting glimpse into the secret workings of life.

Anyone who has ever seen frogs' eggs knows what they look like: small, round like currants, gray and white. Naturally they look different at different stages of life, their own or the beholder's. To a high school boy they look fascinating. Even to an adult they are not completely repulsive. Looking at these "inert glairy masses," as the American Cyclopedia calls them, I have a sense of looking back into the dim beginnings of life, perhaps of man's own life. A high school boy simply looks, night and morning, to see whether they have turned into tadpoles yet.

The pint of eggs, still clinging in their original scooped masses, has settled to the bottom of the jug and lies there motionless except as the activity in the kitchen shakes the jug, stirring the water into slight waves.

At breakfast we discussed how the head and tail should begin to show on the second day, the gills on the third day, and within a week the jug should be full of tadpoles instead of frogs' eggs.

"When they hatch," Dick said, "they must go back to the water hole before they get too big to come out through the throat of the jug. Besides, they must have food."

"What do they eat?" the boy asked in quick pleasure.

"Oh, mud, moss, things at the bottom of the pond."

We estimated the pint would furnish enough frogs to fill a pond, and there were countless more pints in the water-hole.

"Waterfowl eat them, and wild animals eat them, too," said Dick. The adults furnish food for all classes of vertebrae, from fish to man, the clyclopedia says. It is estimated that no more than one out of a thousand lives to reach winter quarters, from which to announce official spring.

I look at the silent, gray and white currants in the jug, and they look like eyes looking back, patiently waiting for their destinies to be unfolded. In the silence, a spectator at this preview of destiny, I imagine myself, invested with authority, leaning forward and gently touching the glass jug with a beckoning forefinger. "You . . . and you . . . and you." And I can almost feel, behind me, a greater and more authoritative forefinger tapping my glass jug and beckoning. "You . . . and you . . . and you."

SHEETS on the clothesline this morning flap in mighty applause. The sound comes in distinctly through the house walls; and, when they flap, reflected sunlight rolls and leaps from the sheets like crumbs being shaken from a Sunday dinner's tablecloth.

The house was cozy as a cup of tea, but when Dick invited me to go with him to the sawmill, I went because I knew there would be spring signs to see.

The furry gray catkins of pussy willows are now covered with powdery gold pollen; they shake it into the wind at every opportunity. Little woods-creeks, which later will go sedately, even dawdle, now run fast, full to boiling over. The green-blue water, where it is clear, makes one thirsty; but where it flows past the fields it is muddy, and sadly you know a few more acres are migrating. In the water of ponds and swamps, the sharp, green knife blades of cattails are already six inches above the surface; and the new green of willows, surely the softest green in the world, can be seen even from a distance.

Along the banks of the road, white hepatica has broken through the dead grass into a long, star-spangled strip. Violets are in bloom, in thick, inviting patches. The long green trumpets of lily of the valley are being quietly unrolled. A few more days and the fragrant white blossoms will be there, like music blown out of them. Oh, April, how gay, how welcome you are! Is this what the poets mean by "April's burgeoning"?

SPRING is not purely a time of joy and unworry. It is also a time of relentless self-assertion and contest in the world of nature.

Animosities that were dormant last winter in the common need now begin to stir and reassert themselves without mercy. The flock of seven Bantams (three roosters and four hens), which spent congenial winter days together and together slept peacefully in the barn through winter nights, have felt the stirring of spring.

Now their happy fraternity is ended.

Yesterday one rooster was discovered cowering bloody and bowed and alone in a side ditch. This morning the brown rooster with the black fountain of tail feathers walks alone, wandering lonely as a cloud. But the little bright orange and gold rooster comes out proudly, leading his harem of four hens triumphantly about the yard, finding a banquet there and imperiously calling them to come and partake of his bounty and his authority.

WHEN THE red-winged blackbirds first return in spring, they sing a melodious song, loud and clear: "Walkee, walkee, walkereee!" Later, when the thrill of homecoming has somewhat subsided, they sing differently. The song is still happy, but sudden and harsh, a two-syllabled, mechanical-sounding cry.

It is, in miniature, the cry of the hay baler when the baler has taken in the last flake required to make a bale and snatches the twines to tie it up. At that ecstatic moment the baler makes an outcry that is both a gulp and a disgorging.

The blackbird's outcry is more scratchy, less rumblous, but the two voices are enough alike that anyone who has helped haul in baled hay on a summer afternoon, with rain threatening, would recognize the similarity. He would even feel the sharp edges of bales stacked in the wagon, and remember the dark, curling clouds in the gathering-gray sky, would in fact feel the sudden lurch of the speeding-up tractor and hear the protest of of the baler, licking up the hot, dry hay from the endless windrow.

The song of the blackbird, like the song of a crow, is one of the songs in which summer is captured and held as on a phonograph record.

"IF YOU READ all the signs," said Dick at breakfast, which is the time we always take for a leisurely conversation, "you realize they work the caretaker with rigid discipline."

Did he mean church, cattle, or children? We had been discussing all three. I waited while he drank the milk out of his glass, poured cream and then coffee into the glass.

"They don't want a lot of loud talk and conversation. They won't tolerate changes. They don't want to be driven; they just want to be served." The church, I decided.

"You try to get them in, and they'll stand in the doorway, blockin' the way. . . ." Yes, the church certainly. His words exactly described the Maple Grove congregation pausing between the old stone wall and the front steps to talk plowing, tractors, crops, and beef prices, reluctant to break it up even when the old upright piano starts clanging: "Bring them in, bring them in." Still, it might be the children; sometimes it's hard to get them in to meals.

"And if one of 'em has horns, she's got things her way. Now horses will go right in and kick the slats out of everything they don't like, but a cow will just stand there, lookin'. . . ."

It was a relief, finally to know.

He dipped a half a doughnut into the coffee glass and went on: "They can make a man wade through briers up to his armpits, or through mud ankle-deep, and here's how they do it." He set a salt shaker in front of the coffee glass. "They get some obstacle between themselves and him, and they watch to see which way he's going. Then they go the other way. But he has to wade through it to head them off."

He finished the doughnut and permitted himself a smug smile. "It's interesting to outwit a cow about where her calf is hid," he said. "A sow hides her pigs and goes off; a mare never lets the colt out of her sight; but a cow will hide her calf and rejoin the herd, and slip back at intervals to nurse the calf.

"You go out to the field, and if you're not watching right then, you'll miss the cue. When she sees you coming, she'll look right straight to where the calf's hid, because the thought of the calf pops into her head before the thought of deception. After that first instant she has her plot laid. You can't drive her, but she'll lead you and you can follow. She'll lead you all over the field, walk right past where the calf's hid, within forty feet of him, and never stop nor slacken her pace. But from that first giveaway glance you know what direction he's in.

"The calf won't make a sound. He won't leave. A cow

and a calf always meet at the spot where she left him. He'd stay there until he died of starvation if something prevented the cow from coming back. If the cow ain't around and the calf ain't one of these wild ones, he may come out and play with you, even go as far as a hundred feet away. But when you leave, he'll go back to the very spot where the cow left him."

ALL WEEK the talk had been of fishing. A simple pole had been readied, with twenty feet of black waterproofed nylon line. "Try and break it," challenged the card on which it was wound. It has an adjustable cork and a hook "guaranteed for twenty-pound test, preferably by bream or trout."

On Thursday frogs chanted. Spring had come to the watery places and fishing was possible.

At supper the conversation was fishy. What better time could providence have induced an expert fisherman to stop here? John Dunning has fished practically every place and practically every way and can tell about it so that your heart aches because you weren't along.

John had eaten supper at home, but he accepted a glass of chocolate milk to drink while we ate hamburgers and cherry pie; and even these seemed almost to have a fish flavor.

John weighs 325 pounds, wears size 54 overalls when he can get them. When they told him at the Dolan country grocery that size 11 was the biggest shoe they had, he asked them why they didn't carry some men's-size shoes.

In school, a one-room country school, John paid little

attention to English and literature, but his acute observation and blueprintlike memory give authority and vividness to his conversation. Nobody questions the absolute accuracy of his fish stories or his fishing lore.

He told us how he once caught a channel catfish with his bare hands. He put a big fist into its mouth, and every time the fish opened its jaws, John shoved his hand further down until his arm was in almost to the elbow. When he drew out his arm, there were toothprints in a row all the way down it.

"Every time he closed his mouth, he raked the hide off," said John, laughing heartily, "but I'd 'a caught him if he'd took my arm off clean to the bone."

Some of your best friends aren't the right people for you to go fishing with. "Ben's a good feller," said John. "High-strung, though. He'll go to pieces and just fly like glass."

Joe told of fishing places he had been invited to come to: Herschel's creek, where Herschel's daughter, Nancy, had seen "big fish" jump out of the water; the little branch behind Carr's barn, where bright little sunfish swam in clear, running water, Carr's son, George, had said; the murky curve of Beanblossom beyond Dick Wampler's barn. John invited him to fish in the pond that had established itself in the field without John's help.

"How do fish get in them places?" demanded John suddenly. "I never put any fish there. But they get in ponds and water holes and rain barrels. I tell you, it rains fish, it does. It rains fish."

Joe showed him the glass jug in which he hoped to

bring home, alive, the fish he caught, to put in our new pond. "Will they live, John, after I take them off the hook?"

"Yes, they'll live if you don't hurt 'em too much, Joe." John is kindly and encouraging, an excellent neighbor. "I've caught fish that had been caught before and got away. I caught a gur-big 'un once. 'Why, he's got one eye out!' I says. 'And his mouth is tore away at the left side,' I says. I hooked my thumbs in his gills and shook him. Like this. 'I'd just like to see him get away now,' I says."

He got his wish. The fish gave him one long look out of its good right eye, flipped over, and was gone.

ON A RAINY NIGHT, sounds sift through the farmhouse walls more distinctly than at other times. For the person listening, they deepen the sense of coziness.

The train goes hurrying through the night with a swishing sound, like something being drawn swiftly through water. Its diesel horn sounds like a cow bawling not far away, although the train actually is miles distant and rapidly getting even more distant. The listener remembers the old-time train whistle, the old-time train smell of coal smoke and tobacco, the old-time harshness of red upholstered seats with a metal corner grip where a hand could steady a small person walking through the aisle. The sound of the train, on a rainy night, enhances one's sense of security. The thought of its people, being borne swiftly toward some distant, pleasant, and surely important destination, is oddly soothing.

Comfortable, one hears the modern sounds: the sound of a plane stitching its way across the dark cloth of the night sky; the rush of cars; the grinding of heavily loaded trucks passing by on the new highway a few miles away. The listener accepts them, as she accepts the nearer, commoner sounds: the tentative whisper of a maple branch against a window; the regular breathing of a child asleep in an adjoining room; the creaking and muttering of an old farmhouse as it stretches its sinews and adjusts itself to a more comfortable position in its sleep.

"Sometimes," remarked the seventeen-year-old son, "I wish I had a broken leg or something, so I could just walk slowly up to the woods—just take my time and see everything, the leaves, the bugs, everything. And enjoy myself."

"Well, goodness," exclaimed I, "you can do that anyway, without having a broken leg!"

"No," he insisted sadly, "if you're all right, you have to be working. You're always in a hurry to get some place else. You don't have time to go slow and enjoy yourself."

When Carol and Joe and I came home from the Library, we discovered Dick's note on the kitchen table: "The kittens are in the loft just beyond where I stuck the fork. Take the flashlight."

With this "treasure map," as Carol called it, we hurried to the barn and up the ladder to the loft. We found the pitchfork marking the spot.

In a dark, hidden crevice, safe between bales of hay, lay the gold-spattered black mother cat, Old Puss, with her five new kittens. Three were a rich yellow, one a dashing black and white, one all white.

"Her traveling man was pale yellow," Dick had said.

Old Puss was glad of our visit and our offerings of roast pork and potatoes. Neither she nor we mentioned the fact that we had been earnestly hunting her hiding place for the last ten days. Now she stretched out her long legs, kneading her claws happily in the clean hay, and let us pick up her kittens and examine each one at leisure.

The gold-spattered cat was, understandably, somewhat smug. She had kept us guessing until the kittens were old enough to be handled, and she was ready to have us help take over the responsibility of their upbringing. Being extremely clever, she let us think we had finally outwitted her.

THE WHOLE HOUSE smells of sassafras.

Yesterday Dick dug up a two-foot length of sassafras root six inches in diameter and last night prepared its bark to send to an ex-Hoosier now living in California.

He washed off the mud and scraped away the dark, thin outer bark, disclosing the white inner bark, which is the thickness of a stick of chewing gum, slightly oily and soft as an underripe avocado. He cut it off in inch-wide strips, whereupon the lovely sassafras fragrance gushed forth.

This would make Indiana's distinctive sassafras tea. Some people put cream and sugar in it, thus making it into "saloop"; most people prefer it with sugar only, in its clear pinkness.

The bare wood, when the bark had been removed, looked remarkably like the bare scraped shank of a freshly scalded hog. Dick laid the bark on a newspaper to dry and swept all the broken bits into the fire. Burning, it gave off a delectable suggestion of what it can do to flavor baked foods.

When stepped on, little bits of bark stuck to our shoe soles and bare feet with gluey wholeheartedness. Tender new shoots of sassafras twig are edible, somewhat glutinous, and in the south are sometimes used to thicken soups.

In Indiana, sassafras is the Abe Martin of trees, homey, rustic, not taken seriously, but appreciated. Farmers don't accredit it highly for lumber but like it for bean poles, chicken roosts, hoghouse braces. It has another unusual property: it is insect repellant. That is why it is good for henhouse roosts. And maybe why pioneer Hoosiers sometimes made bedsteads of it.

THIS WEEK the pink of redbud draws the eye irresistibly to wooded hillsides. This pink is unlike the luminous thin shine of peach petal, or the thick, glossy voluptuousness of magnolia. At first it is almost red, but at its prime, when all the redbud trees stand forth, it is exactly the pink you get when you pour thick country cream over a piece of fresh raspberry cobbler and the juice and cream run together and blend.

Hills that were lately colorless almost to invisibility

now are suddenly like a party table. With a giant spoon you could reach far out into the distant hills and eat this delicious pinkness.

The glossy, heart-shaped leaves, at first almost brown, will not open until the pink flower has had its full day.

As if to accent the redbud, the hills are touched with the thin pale gold of new sassafras leaves and bursts of wild white plum like sudden chords of music. Now can be seen, also, the unbleached-muslin color of immature dogwood. The flowers are small, each petal warped and scorched on one edge as if someone had rescued the dogwood from fire just in time. When the redbud pink has melted away, the dogwood will come into full, rapturous white. It will stand in horizontal flat layers. Then little insignificant bushes nobody had noticed before will be suddenly covered with white glory.

DAYLIGHT comes now around 4 o'clock. While the sky is still dark roosters begin to crow from various farms in the neighborhood. Their voices pierce the dark sky like a knife cutting a design in the top crust of a raspberry pie, and the morning comes through like colored juice out of the slits.

Last summer's brown thrasher has returned. This morning I watched him get breakfast from the ground. He scratched away the dead leaves and grass with the frenzy of a hen in a newly planted lettuce bed; then he drew back his brown head like a hammer and drove his long, sharp bill into the ground with such force that I expected to have to help him pull it out.

He is large as a robin; his body is a thin, bright brown, the color of cocoa in a can where you get too much for the money. His underfeathers are white, striped with distinct, dark wavy lines. He balances himself with a long, rudderlike tail, and everything he does seems done violently.

Barn swallows, too, have returned this week to last summer's old nests in the barn. Farmers consider it good luck for barn swallows to return, even though in their tidiness, wearing deeply forked dark coats and orange-tan vests, they do make an overall-clad farmer look markedly unkempt by contrast.

❧ YESTERDAY, digging up some hollyhocks to reset against the stone wall of the barn, I came upon a creature I was not intended to see for some years yet, a seventeen-year locust, drowsy and supine in its big plastic shell. It had large, bulging, headlight eyes; a flexible, pointed, plastic-tubing body. It showed no pleasure in seeing the light of the extra-fine spring day. It lay on my palm, sluggishly waving the legs of one side of its body. Through the transparent shell dark mottlements were visible, as if it had eaten lumps of earth. It was unbeautiful, certainly, but thought-provoking.

I held it on my palm, remembering how two years ago everyone had wearied of the locust cry, "Pha-roh, Pha-roh," and how we had resented the slashed look in trees where locusts had killed the new growth.

Dropping him on the ground, where he could choose whether to go back to sleep or stay up, now that he had been wakened, I went ahead digging hollyhocks and thinking. A locust has only one summer to spend above ground before he splits his shell down the middle of his back and leaves it. Will he choose this summer, now, in a year tense with the chill of cold war and the lofty coldness of summit meetings, or will he wait ten years and see if that summer is any better? What will that summer be? My belief is that in that summer Indiana farm women will still be transplanting hollyhocks to decorate a barnscape, and little girls will still be picking the long silken blossoms to make into hollyhock ladies, and some creatures that would prefer to be left dormant will still be awakened.

When again we hear that sustained, ancient-sounding locust cry, let it be remembered that the name Pharaoh is associated with one of the earliest and most significant mass migrations toward freedom that the world has recorded.

IT'S NOON and dinner is ready, and if it has to wait it won't be very good.

It's going to have to wait, though, because John stopped twenty minutes ago, and I see Dick has got into the car with him. They are laughing and talking confidentially, and I know when Dick comes in he will tell me, "Oh, John and I did have the finest visit!"

John hasn't been here for quite a while; and since he is a better than average conversationalist, well informed on local news and an excellent analyst thereof, I wouldn't interrupt the visit for the best dinner I can offer.

I could go out and invite John to dinner. How simple and plausible that seems! I'd like to do it; there's plenty of dinner for all of us, and the tablecloth is clean. I'd enjoy having John stay. But if I went out and asked him now, right at noontime, he would say he'd like to, Rachel, but he promised Leota he'd be home. Because that is the diplomatic way to tell a lingering guest it's time for him to go home, and the diplomatic way for him to admit it.

Therefore, dinner will have to wait. A farmer can subsist on a cold meal; a neighbor's visit must be enjoyed at its prime.

THERE IS probably no time when a farmer is more acutely aware of nature's generosity than on that spring morning when finally he turns his long-haired, shedding, winter-fed cattle in on the new grass.

Now the milk flow will increase, the cream be deeper and more yellow on the crock. Barn chores will be at a minimum, leaving the farmer more time to spend in the field.

A sense of deep quiet begins to accumulate in the barn that morning. Swallows remain undisturbed in their apron-pocket nests against the barn rafters. Small mice now can shuttle unmolested from one manure tunnel to another.

Having "made it through to grass," Dick delays the turning-in a week longer, for added protection of the tender pasture sod. Then he makes a note of the date in the farm book, comparing it with last year's record. Thus a man measures the tallness of the year, as he measures the tallness of a growing child and notes it by a penciled line on the frame of the kitchen door.

"EVERYBODY's got 'er on the big wheel today!" exclaimed Dick as I climbed on the tractor to go with him to the field northwest of the woods, where he was going to be plowing for corn. This is a farmer's expression and perfectly describes the joyous frenzy of activity that overtakes farmers on a good farming day in April.

All the neighbors were out that day. Across the road John had his big tractor and his small grandson. They were using a battery drill on the tractor to blow orchard grass on the field because the seed is too light and chaffy to go through the regular drill.

In the next field Ira was plowing for corn. His furrows were dark and straight, as he came toward us.

Dick made happy conversation as we sped along in road gear. "Emsley was out with a walking plow and a team of grays this morning when I went down the bottom road to get the plow points I left to be sharpened yesterday. Edith was picking greens for dinner, but Emsley said he wasn't going to eat any because he wasn't any cow!"

Halfway back across the next farm Carr was finishing up his grass-sowing so Dick could borrow his drill. He had stopped to make some adjustment on the noisy tractor, but saw us and waved.

We turned in at the gap of our north field and crossed the terraced clover field where the two-year old seeding is now a lush green paradise. It was wiry dewberries and ravines when we came. In the next field a one-year seeding promises to match the record.

Beyond the woods line we could look across the green wheat sward and see Earl's hilltop field, where Earl's son, Mark, was plowing. The splut-splut-splut of Mark's tractor came distinctly up to us. The newly turned section of his field was a dark brown stripe next to the pale brown stripe of last week's plowing. The winter-dead grass on the unplowed sod made a gray stripe next to that.

Down the slope toward the east we could see Russell's barn. Russell was cleaning it out with a power manure loader. At short intervals the loader backed out, looking like a long-necked, square-headed dinosaur; then there was a clunk-clunk as the load rattled into the spreader. Beyond the barn, water danced and sparkled in the new pond under the sunlight. "The pond already has enough water in it to last all summer," Dick said admiringly.

Behind Dick's tractor, the plow rolled the mellow earth into little seams as neatly as a foot-hemmer on a sewing machine. All the old cornstalks were folded out of sight in the eight-inch depth of smooth, even furrow. Earthworms came out on top in great numbers, along with the clean, well-rotted masses of grass that had been turned under last year.

"When everything's going well," said Dick, shouting above the singing of the tractor, "you just can't hardly keep from looking back and watching the furrow all the time. It's spellbinding, like water flowing, or music. And you wish you could go on like that forever." That's what it means to "have 'er on the big wheel."

THE POIGNANCY of the Easter season now reaches its highest and most unbearable moment. It is a moment marked by triumph, but closely related, as if indeed they stood side by side, closing a circle, to that last moment of Christmas Eve, in which joy is accompanied by profound sadness.

From the first thrilled moment when the Babe was seen in the manger to this lilied, exultant last moment before Easter daybreak, there has been a steady building-up toward heartbreak and anguish and triumph. It is so every year.

Before daybreak on Easter morning, I came downstairs. It was dark outside. A sharp, bitter wind was running across the farm. I dressed warmly because I was going to an Easter sunrise service in the hilltop clover field in front of our woods. A group of young people from a town church were coming, with their leader, who was a neighbor of ours. They wanted to hold their sunrise service there because of the inspiring view. The Maple Grove church, small, white, old, is just across the road from this pasture.

Sunrise that morning would have been at 5:27, but already it was obvious that there would be no sunrise. The best the sun could do was to give us two pale pink streaks in the east.

Rose followed me up the hill. Two cars had already

come to the gap in the field and were signaling other cars by light taps on their horns. The sound was neither worshipful nor disrespectful, but fearing the service had already begun I whispered to Rose that she must go home. "We're going in very quietly," my neighbor had said. Obediently, but reluctantly, Rose went down to the house.

The services had not begun; the singers had not arrived. The minister was there, shivering in his short, inadequate jacket. Everyone else was shivering, too. On the windswept hill the morning grew. The dark horizon was pierced by staccato points of light: lights from radio towers in town, the glow of town lights thrust up against the dark sky, lights from cars out on the new highway.

Easter services should be at daybreak instead of at sunrise, I thought, shivering, because day always breaks, whether the sun rises or not. Besides, it was dawn when the women came to the Tomb. The darkness of the quiet earth, just before daybreak, make this one of the most poignant and fruitful moments of the whole day.

A minute and a half before the time of sunrise, the services began. The minister spoke briefly; a teen-aged girl sang. It was simple, reverent, impressive in the bitterly cold, sunless morning.

The cattle watched from their side of the fence, their natural curiosity filled to overflowing by the sound of trumpets and singing and the sight of several people assembled. The ponies frankly tossed up their heads and raced across the field. Rose stood in the yard and barked until somebody came out of the house and quieted her. Across the road, the little church was quiet. It was a new and memorable experience for an old farm.

MUSHROOMS do not grow in exactly the same places every year. This spring I found six small morels in the open field near the bee tree. New patches appear in the woods, as if mushroom seed had been washed out in a gentle rain and left there to sprout when the water subsided. In our woods we find snakeheads that have big, hollow stalks and small, pointed, deeply-ridged caps. We find the sponge-textured morels, shaped like Christmas trees, shorter-stemmed and generally preferred to all other kinds of wild mushrooms.

On Monday during Easter vacation Carol and I discovered a new patch with so many we felt selfish about picking them alone. She stood guard "so they won't get away" while I went down to the barn and brought back Dick and Joe.

Our small paper sack was soon filled. (Never tempt a mushroom by taking a large sack; they hide.) Fortunately Dick had his cap on.

Dinnertime came, but Dick said firmly: "No, wait. This don't happen more than once a year."

We searched all through the woods. The children and I always pounce with cries of delight on each new find, but Dick likes to stand and look and casually point out the mushroom for someone else to pick. When he would see mushrooms on his way through the woods on the tractor, he would get off and put up sticks so we could all go back later and pick them. The experienced mushroom hunter knows to look for the mushroom's mate, and also to look under the little humped-up mounds of dry leaves for the mushroom playfully hiding there.

I like the other things we find on a mushroom hunt: the small snails, dark and wet and shiny in the wet leaves, or the delicate white empty shells of others; the woods flowers; the bright-orange shelf fungi on old dead trees and the bone-white suède-textured ones on fallen limbs; the wonderful, loose, fragrant leaf mould; the walnut shells carved by teeth of squirrels and dropped down at the foot of a hollow tree. I like the jokes and family conversation.

Eating a mushroom is pleasant, but the real charm of the mushroom is the conviction you have that it is something created by you, out of nothing except the intensity of your wish to find it. Mushroom hunting is a kind of madness to which one surrenders completely. After each find, however large or numerous, he thinks: "Just one more."

The smell of mushrooms is delicate, but quite distinct. It is somewhat as if a cup had been emptied of pineapple juice and let dry unrinsed out; slightly like smoked wild fowl; slightly like the smell of fresh wet wood. It is a pervasive fragrance and lingers in faint floating wisps about the kitchen after you have split the mushrooms and put them in salt water to soak out the insects. It must be by this fragrance that cattle locate them. They eat them greedily. If they have found some in a place one morning, they will run to that same place as soon as they are turned in to the pasture the next morning.

❧ Coming up from the mailbox this morning with the mail, I noticed that the rapturous pink is beginning to show on the old crooked peach tree at the back steps. Within a week, unless the wind turns unexpectedly cold, the tree will be a pink lamp lighted in the east yard. With a deep conviction that life is good, I sat down, rejoicing, to read the mail.

I saved the letters, like dessert, to the last, and opened the newspaper. One must read the newspaper. It is a way of knowing that the world is doing, whether you like it or not, and of determining your inescapable responsibility in the world. Sometimes the world's doings, as disclosed in the newspaper, are hard to endure.

The front page was filled with such doings: a little boy had been beaten by a schoolteacher; two houses had burned, one with its family in it; a little girl was suffering from an incurable disease. (Somehow these child tragedies seem always to concern children the same age as your own. If you have children, you are always vulnerable. If you are a parent, you are the parent of all children.) Among nations there was nothing but bitterness, accusation, hate. I laid aside the paper, and it seemed to me the peach pinkness had deepened even in the short time that I was not looking. Feeling guilty for noticing this pleasant thing, I looked away from the tree.

But the old tree continued to beckon to me, and I was

aware of it even from the farthermost corner of my eye. It is not a good tree. It is a seedling of some unidentified variety and bears small white peaches in late September. One side of each peach is black-freckled and often the inside is wormy (because we never spray the tree); but the good side is delicious, and the freckled side is even more so.

The tree is only there by accident. It came up from a seed thrown down a long time ago when there was a wood shed there. To survive at all, the seedling had to grow up bending forward, out of the way of the shed, and so it grew into a crooked tree.

Deep in the heart of the seed, as in every seed, was the confident expectation of life and ultimate seed-bearing. Now the shed is torn away, but the old tree remains bent over. Every September it offers its richness. Nobody passes without meeting its hospitality.

Avoiding its pink smile, I opened the first letter. It was from a man in a New York magazine office. He wrote: "New York wouldn't be what it is if it weren't for us outlanders who put away our hoes and hied to this Baghdad-on-the-Subway. You'll find more fugitive farmers in New York City than anywhere else on earth. They wouldn't trade their places here for any other place, but sometimes you see them with that misty look, dreaming of home. That's what makes us go on when the normal frustrations of this monstrous place seem almost unbearable. Sometimes in that loneliness born of a crowd they dream of a more peaceful existence where the skin isn't bruised by concrete and stone. . . ."

Nor the heart by newspaper headlines, perhaps. Maybe

what the world most needs is an old bent-over peach tree at its back steps, offering sweetness, worminess, and comfort, reminding us that we can't solve everything; that it is neither possible nor desirable to eliminate all suffering from life, even from the lives of people most dear to us. Suffering is valuable, having its place in the calculated development of man. This is one of nature's most frequently repeated statements.

The old peach tree, joyously pinkening, reminds us of what hope and determination and simple confidence can do, growing crookedly out of begrudged inches in an obscure place.

MAY

[MAY]

IT was not the kind of rain Forest Payne calls a "goose drownder" and Dick calls a "toad strangler." It was the gentle kind that stops a farmer from plowing and his wife from setting out strawberry plants.

It was a good time to go on an errand to Herschel's. His lane, through the woods, would be burgeoning with wild flowers now. It is a little lane, off a little road, off a slightly larger road.

To get there, we passed Earl's pasture, where papaw groves gave brown-red promise of a good crop; passed Mark's hillside of quaking aspens now with tiny leaves and tassely blossoms hanging down like long earrings; passed Scherschel's ravine, which is deep as an orange cut almost through to the peeling on the other side; passed Walter Hasting's neat little house at the foot of the hill; and so came, a short distance beyond, to where Herschel's unconquered long driveway starts in the woods and toils up at the easiest level, like a cow climbing a hill, to Herschel's solid old farmhouse.

At the rutted entrance of the lane, wild cress set the pace in a golden mass and the white blossoms of wild strawberries followed. There, in that dense, flower-brimmed woods where leaf mould is ages deep, the year has spent all its winter savings to create a lavish May moment of flower and leaf.

New leaf buds on hickory saplings were red and shaped like orchids. Heavy fronds of winter fern, now bronze-colored and dying, lay on the banks; and from their hearts new fronds have risen, tall, white-green, with small, fuzzy, curled-up tips like snail shells. They are called fiddleheads. Maidenhair fern had chaplets of new green leaves on wiry, black stems; sweet fern was there, with new leaves symmetrical as the rhymed lines of a poem. Fragrant sweet William, in varying shades of lilac, bent under the burden of fresh rain. Straw-colored, long, frayed-out bellwort hung from tall, pale stems.

Almost every woods flower was there: white trillium and its sister, red wake-robin; yellow adder's-tongue; May apple in full open bloom like a water lily; Indian turnip with green-and-white-striped canopy; wild geranium with ivy-shaped leaves and lacy buds; small, deeply blue bells of Jacob's ladder; white violets with long, narrow, heart-shaped leaves; yellow violets bright as gold coins; purple violets, and blue ones, and violets striped blue and white. There was a tiny white flower with ten blunt-ended petals like spokes in a rimless wheel; silken, yellow buttercups; bloodroot; spring beauties; squirrel corn not quite in bloom; wild ginger in a thickening mat that tried to hide the red flowers underneath; new ginseng plants; beside a rotting log, two vivid scarlet-cup mushrooms.

On the way home we saw two quails quietly sitting under a hackberry root along the east road. Last year's hackberry stems were like knots of black number-eight thread. The new leaves will be warted underneath, with green warts where these were black. "A person really needs a buggy to see everything," mourned Dick, driving as slowly as the car could go and still be moving.

❦ ALMOST any farmer can describe blackberry winter.

It's that cold spell that comes in May, about three weeks after spring fever. It comes when blackberries are in bloom and does sometimes actually drop a few real snowflakes into the white flowers. It doesn't bite through to the hard, green, incipient berries nestled behind the petals. It lasts less than a week.

It is earnestly cold. A farmer plowing a cornfield in blackberry winter wears the flaps down on his winter cap and puts on all the coats he wore through real winter.

Most farmers will not give an exact date for the beginning of blackberry winter. But the shoe repairman, who used to live on a farm and still has a farm on which he keeps sheep, didn't hesitate when I asked him.

"May 10," he said.

Most farmers will agree that it's "around" that time. And most will agree also that its over-all accomplishment is to make people thankful that July is what it is.

I WENT UP to the woods yesterday to see about some of my good friends that live there, especially the old beech tree at the south edge of the woods.

Its circumference was almost three times the length of my stretched-most scarf. The roots that anchor it solidly rise like strong muscles above the surface of the ground.

Two boards, nailed to its trunk long ago by someone who had reason to climb it, are still there; and one side of the trunk is scarred from someone's target practice. These marks of human association only add to the authenticity of it; a beech tree never seems fully grown until it has initials carved on it, or other evidence of human use.

At the height of a silo the top has been broken out, in some wild windstorm, but new wood and bark have developed around the dark outline of its hollowness and sap is still carried to its top branches. Beech trees become hollow and storm-vulnerable before anybody knows it, except squirrels and birds, to whom the hollowness is good fortune.

The blacksmith shop

The sharply pointed leaf buds are bronze-green and tightly rolled. They do not conceal the gnarled, stiff profile of the tree. Its character is revealed in a blunt honesty. It has borne its blows without self-pity or bitterness. What's more, it has achieved a look of great serenity in its gray, fine, knitted-texture bark. Where there are no scars, the bark is smooth as water flowing, smooth as a fish's side. The grayness has vigor; it gives comfort. The tree seems like a person.

"How have you fared through the winter, my dear old friend?"

"I have lived, thank you, and I have found life good."

"I HAVE TO take the plowpoints to be sharpened," said Dick, "and I want you to come with me. I never saw so much dogwood in my whole life as there is along that back road."

The blacksmith shop is on Walter Parks's farm twelve miles from town. A rough, small shed, it stands in the sun-bitten corner where his driveway meets the road.

His two sons, aged ten and twelve, are expected to help with the work when they are not in school. The oldest one was pounding the rusty iron teeth out of an old harrow and seeming to enjoy it. Dick had brought them a banty rooster. They were doubtful that their father would let them keep it, but he did.

Mr. Parks didn't think much of the previous sharpening job on Dick's plowpoints. "It was electric-done," he said, "and either he don't know how to sharpen points or else he don't care." Plowpoints are not sharpened like knives; they have to be heated and beaten out to a point. Some smiths use an electric trip-hammer, but Mr. Parks doesn't have electricity.

An assortment of machinery and piles of machinery parts were scattered over the floor. Smoke rose languidly from a layer of coal on the forge.

Mr. Parks starts work early but takes the day off on Saturday to go to town.

He showed us a pair of points for an old Oliver walking plow and said if anybody could use them he would sell them cheap; he wished he knew where he could get a hatching of goose eggs, but not from yearling geese because he thinks their eggs don't hatch well.

A couple of years ago, he said, he invented a wind-up fence stretcher and could have sold the patent rights to a neighbor for $700. "But I just never did go ahead and get the patent," he added. He has been blacksmithing for a good many years, twenty of them for a circus. He lives on a farm because he likes to, and keeps the blacksmith shop so he can live on the farm.

The land in that dogwood-glorified community is productive when well cared for and protected, but the tillable areas are in small patches. Unless it is protected by a cover crop, the soil erodes quickly after being plowed. On one side of the road you may see a fine clover and alfalfa sod and just acoss the road from it a field gashed as by a giant currycomb.

On the way home through the white wilderness of dogwood we passed one starved farm. Dick noted the sagging barn, the despairing fields, and remarked, in genuine sympathy: "That poor fellow's in bad shape if his wife ain't working."

❦ SPRING PLOWING is a ritual for most farmers, even though they may not realize it. There is a genuine physical pleasure in the manipulation of the soil. When I plant the vegetable garden, I always want to take off my shoes and walk barefooted on the freshly worked-down soil, sun-warmed and yielding underfoot.

Man's love of land rises from more than the need of food and shelter and enough cash to pay the taxes. It includes the passionate love of spaciousness.

Even a nonfarming passerby feels a sense of security from seeing farm land that is well cared for, productive. It is because he realizes that the prosperity of a country depends largely on how its soil is used. History provides bitter examples of civilizations that perished because their people neglected soil and water resources.

A person working daily with the soil is made solemnly aware of his responsibility for it and his obligation to it.

"The destiny of the whole creative process is in our keeping," says a soil conservationist. "The productive topsoil is all we have to live on—trees, grasses, flowers, beasts, and men. When the topsoil goes, we shall all go."

The conservationists do not intend, however, that the topsoil shall go. "You know the methods," reminds one of them, "crop rotation, terracing, strip farming and reforestation, permanent grass and sodded waterways, water conservation, winter cover."

❦ A HORNET had got into the dining room, and now wanted out. Frustrated by the transparency through which he could see his familiar world but not reach it, he threw himself against the glass windowpane with a loudness as if a bird had struck it.

He continually slid down the window and struggled up again. He kept rubbing his front legs with a friction paste that he secretes in his mouth, to gain a better foothold. His body seemed made of thin yellow steel springs dipped into a thin coating of black and yellow plastic.

As he struggled, he expressed his opinion of the whole place in a low, unlyrical monotone.

The only way to show hospitality to such a guest is to capture and carry him outside. I used a jelly glass and a post card.

I carried him out to the porch; when I slid the card away from the jelly glass, the hornet was out immediately. His wings took hold so quickly in their familiar medium that they seemed more to exult him into the air than to propel him into it. He was instantly gone, leaving me to realize what a beautiful thing freedom is. The freedom of any creature is so joyous a possession that you can almost see it in the air, like the transparent, quivering current above a hot stove. To any creature freedom seems more precious than life itself.

A good word for burdock

❦ IN JOSEPH MEYERS'S *Herbalist*, burdock is listed as a biennial, which means it works in two-year shifts. A plant blooms limitedly the first year under the sponsorship of its adult leader while sprouting in the ground getting ready to be next year's beginners are the seeds of last summer's burrs. The two-year-old burdock, meantime, is having the best years of its life. After that, it doesn't amount to much; but it has had a good life, anyway, unless sheep found it. They will eagerly eat it into the ground. Cattle leave it strictly alone.

"Coarse," the *Herbalist* says of this Arctium lappa, not intending to be snobbish, for the *Herbalist* can find some good word to say for nearly any plant. "Obnoxious," say farmers, and mean it.

The rounded flowers, like clipped lavender tassels turned upside down, are enclosed in a pale-green case of gentle-looking threads that later become tenacious hooks. Blossoms grow in clusters. Even then, pressed together, they cling tenaciously; and little girls make baskets or purses with handles from them.

The *Herbalist* says that burdock is "diaphoretic, diuretic, alternative, aperient, and depurative"; that its leaves are "cordate, oblong, dentate, rough, petiolate." Also, they are large, odorous, shaped like the leaves of yellow violet, and start wilting as soon as broken off the plant. Their surface is not as blankety as mullein but will suffice for stick-doll blankets, in a hay field. They are also useful for carrying home wild berries or snails or interesting beetles picked up in a field, or to line a hat to carry these things in.

But any good word you want to say for burdock must be said before the burrs ripen. After that, peaceful coexistence with it is impossible.

❦ So I CAME to the apple tree in the fencerow, now in full bloom. The sight thereof was like a blow, and it was necessary to sit down on a clovered terrace and rest from it. Bees had already discovered the glory of the tree: one bumblebee fumbled at a white blossom, and there were so many honeybees busy there that they made a stream of sound like the noise of distant, rushing water. The apple tree will need them all to get its unnumbered thousands of pink-white flowers pollinated.

For the tree is utterly in bloom. It is white with flowers. Only when I came near enough to sniff deep into a single blossom was I aware of the pale, pale pinkness in the opened petals, but the unopened buds were richly pink. The blossoms hang in clusters, and the clusters are as big as apples.

The boughs bend toward the ground; the lowest come within a few inches of touching it. They bend as if by habit, from many years of bending down with their jeweled burdens of apples. The density of bloom does not quite hide the small new leaves at the other end of the stem, nor the curve and darkness of the tree boughs, nor the piece of rusted wire fence still fastened to the tree, which, being so convenient in the fencerow, once served as a fence post. This is symbiosis between fence and tree.

The glory of the apple tree

It is one of God's special uses for fence that it provide shelter and survival for weeds, briers, and young trees, as well as for birds and small wild animals. Long ago an apple seed sprouted at the fencerow, and because its straight young trunk was able to hold a fence "steeple" firmly, it was allowed to stay. Now the wire fence is rusty and broken, but the tree is in its best years. The measure of its widest expanse requires sixteen of my longest steps.

Last fall, searching for wild grapes, I had discovered the last of the apples. They were dark red, firm, juice-filled. Bushels had already fallen on the ground and spoiled in a winy fragrance, to the delight of many brown wasps. Enough still remained to combine with wild grape into a jelly that was the equal of black raspberry.

An apple tree does not have to justify its existence by bearing fruit. Its fragrance, so delicate that it is almost stronger in memory than in reality, is sufficient. Or the sight of it. Irresistibly you step close enough to inhale from the heart of one bloom, although actually the fragrance is more distinct if you stand back a few steps letting the sun-touched wind bring the perfume to you. No one can ever forget the smell or the sight of a wide-spreading apple tree in full bloom.

Nothing in the whole farmscape that morning could compete with the apple tree, neither the luscious pink of red-bud in the woods nor the quiet splendor of raw gold on the tall, sprangly sassafras bushes further down the fencerow. It was as if they all stood ungrudgingly back that morning, letting all the attention be turned to the apple tree, giving it the full measure of its hour of glory.

A BLUE JAY's feather had fallen in the yard, and I picked it up marveling at the way a blue jay succeeds in combining blue, white, and black into its splendid shade of blue.

It was a tail feather, five inches long, flexible, so intricately formed that it seemed simple. The quill end, which had held the shaft in place in the bird's outer skin, was white. Beyond that the shaft began to darken, and an inch before it reached the arc of white, at the tip of the feather, it had become almost invisible between the areas of dark fibers.

The two halves of the feather are not identical, either in size, color, or angle. The narrower half is basically blue, cross-marked with short black hatchmarks. The wider part blends so subtly into black that one cannot possibly say, "right here is where the blue ends and the black begins."

The effect of this subtle color-blending is that the blue jay's feather seems to change color when turned at different angles. It is like those old-fashioned advertising signs that mentioned men's suits as you approached slowly by horse and buggy, and mentioned something else when you looked back after having passed the sign.

The blue jay's feather has the textured, dry, luxurious sheen of taffeta. This taffeta is made up of a single layer of evenly placed strong fibers (properly called processes) laid side by side. Not one fiber overlaps another; yet normally there is no space between any two fibers—that would make a slit in the feather. You can pull them apart, but you can also run a finger along the taffeta, stretching it without pulling the fibers apart. This strength and expandability is achieved by means of tiny hooks and fasteners (called hamuli and barbules), which fasten the fibers together and enable the blue jay to keep the feather intact against the pressure of air during flights, or when the wind ruffles them at any time.

Holding this tiny quill tip between my thumb and forefinger, I tried to imagine writing with a pen made of it. The quill end was less than an inch long, hardly larger in diameter than a pencil's lead. I could not feel any pressure from its weight against my skin, and I wondered whether the loss of this weightless, commonplace, marvelous feather would affect the blue jay's balance in flight.

Of course, as soon as a feather fell, the glands of the blue jay's skin would set to work developing another feather; and in the meantime the bird was probably too busy gathering worms for his young family to do much fancy flying, anyway.

❦ IN THE CORNFIELD this morning we had to move a log out of the way and thereby disclosed a deer mouse's nest, with the mouse and five mouse babies in it.

The mother's first reaction, to which she instantly gave way, was self-preservation. She ran a short distance; then her maternal instincts wakened and she stopped. She paused briefly, either to gather her wits or to compose a small mouse prayer, then ran back to the babies in the nest. She made two quick circuits around the inside of the nest, and by the end of the second revolution the mice had attached themselves to her.

"They've just hooked on wherever they could," exclaimed Dick.

As she ran out into the surrounding tall grass, they were carried along, bouncing. In size they compared to her about the way a bunch of good young weanling pigs would compare with a small sow. As she ran, the grass and weeds kept brushing the baby mice back so that just as she disappeared from sight they were all hanging, like tassels, from her hips.

"Well, I do know!" exclaimed Dick, admiringly.

EVERY YEAR I think there never were so many maple seeds before, and every year it is true. This year it is spectacularly true. The crushed-stone driveway is brown with them, like a woodshop floor with wood shavings. On the mowed grass where the maple "keys" have fallen, flat wing-end up, the yard in daytime has a look of fur. At evening it looks as if touched by an early November frost.

Myriads have fallen; but the trees have not nearly unburdened themselves yet. The browning pods hang in clusters, four to twelve in a cluster, giving the trees a scorched look as if a bonfire had got too high under them.

The pods fall steadily all day, all night, with the slow continuity of gentle rain. To watch their fall is a pleasure.

The light, flat, ribbed wing falls slowly, whirling, spiraling down so that the heavier, seed-bearing end touches earth first, thus getting whatever moisture, earth-warmth, and cover is to be had.

Falling, the pods brush the house walls with a dry papery whisper; or they strike each other in falling, with a soft, wooden clash. This is a dinner-bell sound to chipmunks. They bite off the hulls from the seeds and stuff their cheeks until they bulge from just below the eyes to just above the shoulders.

Rain carries away long drifts of seed pods to distant creeks. Birds eat many. Night-feeding animals feast on them. Insects sting some, destroying the seed within. Yet the surplus remains.

The solemn purpose of these small capsules of life is attested in the hundreds of tiny, exquisite new maples sprouting suddenly from last year's seeds. If only one seed out of each million of this year's seeds survived, we should soon be enclosed by a maple wilderness.

Such tremendous potential of life and determination to live is almost frightening. It would be terrifying, in fact, if it were not so reassuring. It is nature's succinct way of saying the earth shall not be made desolate; and in these angry, uncertain times, one is glad to know that.

❦ SOMEBODY laid an armload of gunny sacks across a ladder rung in the barn, and a wren built her nest in the dark security of the folds. The round, thick nest walls were lined with downy feathers from her own small, brown body. Now the nest contains seven heavily-freckled, grape-sized eggs. Seven is a large family for so small a bird, even though the wren is hard-working.

The two starlings that built nests in the deep hollows of concrete blocks in the barn walls did not undertake so many. Four eggs have hatched in each nest. If you bend above these nests and make even the sound of breathing, four enormous yellow-lined mouths fly open instantly, wide as a suitcase, in each nest. The starling mother brings only insects, never cherries or strawberries.

Two barn swallows returned this year to last year's pocket-shaped stick and mud nests against the rafters of the barn-loft floor. The swallows fly off the nests whenever a stranger comes into the barn, although their nests are too high to be looked into. The swallows return every year on almost the same date in April, a small fact that gives a sense of pleasant establishedness to an old barn.

When I was leaving the barn yesterday, I looked back by chance and discovered a Bantam hen sitting secretly on a stolen-out nest in the hollow between two bales of hay at the front of the barn loft. Nobody else had seen the nest. The hen looked back at me serenely. Perhaps she realized that we both knew she has set on the eggs too long for them to be of any value if taken out of the nest now.

The barn is also enhanced by babies other than birds. A red sow lying in a quiet corner was surrounded by eight charming, copper-colored, fat pigs. Their velvety flopped-over ears are as clean as if freshly washed for school. The copper-colored hair on their solid tube-shaped bodies looks pleasantly soft to touch and gives off a coppery glint in the sunlight. If you picked up a pig carefully by its hind leg so it would not squeal and stir up the sow, he would feel, as farmers say, "hard as a darnick."

Carol's riding mare has a baby, too, foaled in the barnlot day before yesterday. While the baby walked unsteadily around the mare hunting the food she instinctively knew was nearby, the pony mares all lined up on the other side of the fence, watching with the tender solicitude and rapt interest of neighbors admiring a new baby and giving advice and compliments.

At the barn these May days, life is young and tender.

❦ THIS IS GOING to be a hard week for Polly, the leaf-chewing brown wasp that is trying to build a nest inside the mailbox. For this week has been franked by the U.S. Post Office Department as the week to glamorize farm mailboxes. The rural mailbox is one of the most vital parts of farm life and, therefore, Polly must go.

Every year this happens. After some weeks of daily tearing away her day's laborious accumulation of gray paper cells, I finally have to wipe out Polly. It saddens me, because I dislike to kill things; but it would probably sadden the mailman more if Polly stung his hand.

I have watched the brown wasps (genus Polistes) build.

Theirs is a wonderful, laborious craftsmanship. They bring a wad of chewed-up leaf or old wood to the nest. Hanging upside down, clinging to the nest with one pair of legs, they use the other pair and their mandibles to knead the insipid wad of building material for nearly an hour. When it is finally a thin strip, properly rolled up into a ball, they bite it into place against the cells already built and suspended by a single paper stalk. This takes fifteen minutes more.

I regret to destroy such patient, beautiful work. It is mysterious how every year one Polly returns. It cannot be the same one, because I have to kill that one every year. Yet every year only one wasp returns to this rather peculiar place. How? Why?

❦ AT NOON when Dick came down from the cornfield he brought his annual offering of bloom, bud, and leaf from the tulip poplar at the edge of the woods.

The tulip blossoms are pale green, gently rounded at the base. The six petals start out to be creamy white, put on a jagged wide strip of orange, and then green sets in irregularly, reaching to the gently turned-back cuffs of the petals, so that all in all the blossom is a pale green cup with a flame in it.

The pointed core that later bears the winged seeds is green now, too, and surrounded by flat white sepals, pollen-dusted. In an imaginative mood one could see them as smoke rising from the flame.

On the branch Dick brought today the long, green buds

were hard and pointed. "That's the size they are when squirrels climb the tree to eat the buds," he said.

The tulip leaves are heavy, glossy, in outline somewhat like a child's drawing of a tulip. They flare out in four points and are shaped, finally, like a tulip flower a child might cut out of glossy green paper.

Every year it is a fresh delight to know that a tree bears flowers like a garden plant. On the other hand, you might expect almost any remarkable and lovely achievement from the tulip poplar; it is Indiana's state tree.

❦ LOCUST TREES are in bloom now, too. The smell of them is wafted all over the farm from just half a dozen trees in the corner of the south yard. Their fragrance is more outspreading than squirrel corn, which it somewhat resembles, but not as heavily sweet as the fragrance of tuberose or milkweed. You could live all day with the fragrance of locust and still love it.

A brittle branch brought inside will quickly involve the whole house in a pleasant, fresh perfume. It will also quickly droop and wilt. The pale ovals of locust's multiple leaves are not for house bouquets. The dingy-gray twigs are brittle, angular, and easily broken off.

From a distance the locust tree has a foamy green whiteness. By daytime it is so sought out by bees that the trees themselves seem to be softly humming.

This is black locust, which makes little drab pods. It is a legume and gathers nitrogen in the soil, enriching it, but is reluctant ever to give up the soil it has enriched. If you chop into a locust root, a tree will sprout there.

❦ IF A farm woman ever wondered why she wanted to live on a farm anyway, the answer is here today.

Today the tamarisk bush is a bowl of pink foam; the sweet shrub offers its dark red knots to be tied into a handkerchief for all-day fragrance. Lilacs lean toward the driveway, extending their purple and lilac clusters and reminding everyone that the end of school is near, with pomp and circumstance coming up at three levels of Commencements. From early morning until sunset newly opened leaves experiment with shadow patterns against the weathered boards of an old shed's wall. The air is as comforting as a cushion leaned back against.

Now little boys boast of the number and size of snakes they have seen, and marbles and kites have been forgotten. Fires have gone out in the farmhouses.

Wild lettuce, lamb's-quarters, dandelions, and black mustard are big enough for greens. By suppertime they will be cooking in the big kettle, filling the kitchen with an irresistible summons. There will be boiled beans and cottage cheese, corn bread and apple pie to go with them.

It is a time of ambition greater than any possible fulfillment; of rapture beyond any reasonable, material reason. The joy of having lived to see this day is immeasurable. Some day next winter, when cold and mud and despair threaten to engulf her, let the farmwife remember this day. In order to make sure of remembering it, she can make a note on the calendar: "I liked this day."

❦ ON THE concrete floor of the front porch this morning a snail has traveled and written a record of its trip, a wavering line, nearly eighteen feet in all, crooked and silvery and hardly a third as wide as a typewriter ribbon. Under strong sunlight the record gleams as silver would under the assault of strong light.

The writer was a dingy bone-gray color, no bigger than a dime, and came up from a scraperload of earth dumped yesterday at the front edge of the porch.

Snails are moist and glutenous inside, and harmless, even when addressed by their Latin name, Helix albolabris.

The snail gets around by expanding and contracting his one, large, ventral foot. He travels slowly, stepping on his own slimy trail. He breathes through gills, has eyes on the ends of his antennae which he draws in unsociably when he meets a stranger. Sensitive to cold, a snail seals his shell in October with a hardening secretion from his own body. Late in spring the females deposit thirty to a hundred eggs, which hatch twenty to thirty days later. The young snails are only one and a half whorls big. Ordinarily snails eat only vegetation, gathering it mostly at night. But again, when they feel in the mood for it, they eat each other. You will always find some on a trip in the woods because they like woods and moist places, although, if necessary, they can last through long dry spells. They are sometimes eaten by people; but their greatest contribution has been to the language of humans—the traditional "snail's pace."

THE SWEETEST TASTE of freedom is the first sip.

It comes next Thursday, for school is all but out now. Next Thursday will be a whole unaccustomed free day set between unimportant school sessions. One on Wednesday, when the die is irrevocably cast; and an even shorter, less important one on Friday, to get the report cards.

No day in the whole summer will ever be so delicious and free as that Thursday, for after that, freedom will be taken for granted and its flavor will not be quite so excitingly sharp.

THE ULTIMATE harvest of a farm is not entirely in the crop or income it produces, but in the background of impressions it gives to children who have lived or visited on it, and in the way they remember it, are later influenced by memories of it.

How, then, will this farm influence the behavior of two-year-old David, son of a former Hoosier who is now a minister in Georgia? David's father was to be the guest speaker at the Maple Grove church on the evening of Mother's Day, but David was too weary to sit still for two consecutive seconds. Neither David nor his parents had ever visited this church before.

His mother, pretty and distressed, was sitting in front

of me. When I suggested my taking David home with me until after church, she stood up immediately, gratefully. She picked up the little boy, and we went out to the car. "Let's go see the dog," I invited him.

"Good-by, David darling." His mother put him quickly into my arms and went back into the church. The little boy was suddenly as still as an empty chair. When I put him into the car, I had to bend his legs for him, so he could sit down. I drove away quickly, hoping he was not in shock.

Rose came to meet us in the fading dusk, and I was thankful for her unfailing interest in all children. David looked at her solemnly and touched her black and white coat. She was warm and loving and smelled vilely of the barnyard perfume she loves to roll in.

The road up to the barn was barely discernible. I picked up David and carried him toward the looming dark barn, and spoke gently to him.

"This is the farm, David. This is the barn." He made no response. The white cattle were almost invisible in the dark pasture; the red cattle were only lumps of darkness. We stopped at a barn window.

"This is the pony, David." But the little dark mare and her colt were only more darkness in the dark stall.

We went on, and I opened the Bantams' door. "These are the chickens." From their perch the Bantams obliged with a sleepy murmur, but they, too, were invisible. We went through tall grass to the pasture where the sorrel and white stallion paced restlessly, tethered by a long chain. David neither spoke nor moved. Frogs chanted from the pond; a whippoorwill sang; a cow called a low, imperative message to her calf. The May night was sweet with the poignant fragrance of black locust bloom. All these sounds, the voice speaking gently to him, and the arms holding him were unfamiliar to the little boy. He leaned against me, unprotesting but completely still.

As we repassed the barn, starting down to the house, he spoke suddenly: "Cow, cow."

"Yes," I agreed, and hugged him.

We repeated this conversation until we arrived at the kitchen, where David ate a roll, first gravely shaking off the jelly from it. Once he said gravely, "More," and once, "Rose."

He would have gone to sleep afterward, but guests came and we sat out on the front porch, talking, until church was over. When David's parents came, he recognized his mother and leaned toward her, but not as if trying to escape.

What will David remember of the friendly dark night; the farm noises; the locust fragrance; the talk and laughter of friends in the dusk, who were strangers to him?

Perhaps someday the elusive, unidentified memory of it will help him feel that the dark unknown is not necessarily unfriendly.

No matter what date the calendar says, summer is confirmed only when the whippoorwill has sung.

This morning the early dark hour belonged to the whippoorwill. He had already sung from the hilltop at bedtime last evening. Just after sunset and just before daybreak are his chosen times for singing. To some people his is a lonely, saddening cry; but for other listeners it is poignant rather than sad. It expresses an elusive yearning, as if some indescribably lovely thing were forever just within reach, but never quite attained.

When the whippoorwill sang this morning, he sang earnestly, as if he had a message he felt he simply had to get told to the world. I think he could not have been sitting still in one place while he sang. The sound increased and diminished as the sound of a bell does when the bell swings back and forth. His voice traced a circle, as if he were walking around or turning his body around and around to make sure everybody heard.

He is called whippoorwill because people have always heard those words in his haunting cry. Sometimes, though, the words sound otherwise; often he seems to be asking: "Wherefore live? Wherefore live?" Last night it sounded as if he were urging: "Pray for him, pray for him." He sang with the zeal of an evangelist and "him" could have meant anybody you happened to feel was in need of prayer.

The shy whippoorwill seldom lets himself be seen. If you do catch a glimpse of him, it is usually in connection with an incident that makes you remember.

There was one night when my sister Nina had come down to stay at the farm because I had told her she could almost certainly hear the song, which we both love and remember from a country childhood.

The whippoorwill did sing, from the hilltop near the

church. We stayed out in the chilly wet air listening after the rest of the family had gone to bed, then impulsively decided to drive up to the church to hear the voice closer.

Halfway up the hill the car's headlights picked up the intense gleam of eyes on the road. I stopped the car, and Nina got out. "Why, it's a bird," she cried softly and walked to within eighteen inches of it. The bird sat as still as if it were unable to move. She could see the short neck, the striped feathers, the bristled bill, the intense huddled stillness of it.

She said, very softly: "Come and look." But I thought she had said: "Turn off the light"; and did so for only the barest flicker of time. That was enough. When the lights came on again, the bird had vanished. He had made not the slightest rustle of sound in going. We drove on up to the church, waited a few minutes, and then started back down the hill again. And heard again, this time from much farther back in the field, the same, pure, haunting cry: "Whippoorwill, whippoorwill!"

TODAY we picked the first rose.

Roses are symbolic of summer. Probably nothing represents the idyllic conception of summer tranquility more completely than the old-fashioned pink rose, fully opened, sunwarmed. Summer is never actually as tranquil, eternal, and changeless as the rose believes. Summer is a wind-driven, rain-shaken season, with lines of conflicting activity blown crisscross and into each other: summer vacations shuffled like a deck of cards and dealt out to the fortunate; trips to be made to cooler places and rested up from; gardens to be sprayed, dusted, cultivated, abandoned, picked, frozen, eaten, canned; 4-H meetings to be met and conquered; crops to be fought for and protected; sudden, heavy thunderstorms to be got out of; life to be lived wholly.

The serene philosophy of the pink rose is steadying. Its fragrant, delicate petals open fully and are ready to fall, without regret or disillusion, after only a day in the sun. It is so every summer. One can almost hear their pink, fragrant murmur as they settle down upon the grass: "Summer, summer, it will always be summer."

JUNE

[JUNE]

❀ ❀ ❀ ❀ THE early note of excitement is gone from the frogs' chant now that the prelude is over. At night now their chant is slow, steady, expressing contentment. It goes in the balanced rhythm you hear when a cow is being milked by experienced hands—one-two, one-two, straight into the bucket, while a thick, silencing layer of foam builds up along the outflaring sides and rises toward the bucket's thick rim.

Frogs mark an important notch in the farm year, and in the education of man. In elementary zoology this spring students are dissecting frogs, working from there to a better understanding of larger life. By the time a farm boy gets into zoology, he has already done a good deal of preliminary research on frogs.

"Do you know you can tame frogs?" Joe asked me the summer he was twelve. "I have done it lots of times. If you pet a frog behind the eyes, you can set him down and go away and he will stay there until you come back."

It takes about fifty strokes, he said, during which the frog just sits and glares out of his bulging, periscopic eyes, sometimes lowering them to the level of his head.

My personal belief is that the frog was not actually tamed so much as lost in contemplation of this wonderful creature called boy. That I could understand, having marveled at it myself.

EVERY FARM on which people have grown from small children to teen-agers has its own little place for burying the beloved farm pets and the small, pitied dead picked up from field or roadside.

On this farm it is down in the corner of the front yard, under a couple of locust trees. Flat rocks and slabs of wood mark the graves of banties and kittens and other lamented friends; one gorgeous red cardinal whose death convinced the children there has to be a heaven for wild birds; the crippled, convalescent pigeon, killed in its hospital crate by a stray cat unwisely trusted on the ground that she had herself been befriended; the small, ill-fated bluebird whose nest was torn apart and scattered out of the tree; one doll.

There were others. This morning Carol and I added a naked, dead bird we found in the driveway. The stiff stubs of its incipient wing feathers suggested they would have been blue. We wrapped it in a burdock leaf and committed it to the compassionate earth, realizing it was a slightly childish thing to do, but feeling better about it, anyway.

❦ HERE IT IS, only June, but already a 4-H'er is hard put to find enough unbitten, perfect leaves to make up his collection for first-year forestry.

❦ WILD IRIS, also known as sweet flags, were in bloom yesterday when I drove home along Beanblossom road. They thrive in boggy side ditches and swamps where the water stands two to six inches deep over acid, puttylike soil.

The lavender-blue flower has three curled-back, orchid-like sepals, dark-veined and rising from a yellow base. On top of each, like a small apron, rests a paddle-shaped small lavender petal. The narrow, pointed buds are darker than the opened bloom, which as on garden iris lasts only a short time and then shrivels into a clammy wad. The wild-iris wad hangs from a long, ribbed pod containing small white immature seeds set in rows like beads. As one flower dies, another opens, so that wild iris blooms continuously from May to July.

Thinking it would be nice to transplant some of this aquatic wild plant into the swampy springlot here, I got a shovel and went back to the bottom road to dig some. It had looked as if it would be simple to stand on the dry edge of the road and lean out and lift out a shovelful of the iris rhizomes. I received a great surprise when I tried it. The shovel cut readily through the roots, but lifting them out was another matter entirely. It was as if I were trying to lift the weight of the whole road. I had to abandon the shovel, and all effort to keep my feet dry, and get down with determined bare hands and pull out the rhizomes.

There is good reason for this tenacious suction, of course. If wild iris were as easily lifted out as garden iris, the spring floods gushing along side ditches would carry out all the plants to wherever Beanblossom's muddy waters finally go.

Nature has an admirable way of providing her plant children with the talent they need for survival and then casually leaving it up to them to survive, expecting no further laudations from her.

COMMENCEMENT EXERCISES were held in a maple tree here Sunday evening, just before dark, for two spring-hatched blue jays. Four adults watching from the tree provided pomp and circumstance as the young birds walked down the side of the tree, stepping quickly but unsteadily, as if on unaccustomed high heels, in the bark crevices.

On the ground they separated: one ran east toward the driveway, one went west to grow up with the country. The four watching, applauding adults flew in four different directions, lending encouragement.

The westbound graduate hovered timidly at the base of a tree, then hopped a few steps on the cold, damp ground. His posture plainly expressed every hope and fear he felt. He crouched, obviously wanting to fly but afraid to try it. The parents flew from one low limb to another, giving demonstrations, and the young bird watched them, his eyes lifted up and full of longing.

He crouched again, finally spread out his untrained wings and rose. It was only a short flight and not higher than ten inches above the ground; but it was flight, and obviously the taste of it was thrilling to him. He ran a few steps farther, crouched, hesitated, crouched again, and finally rose again, this time higher and farther. Watching him, I could feel the longing and pride of those rumply-feathered, humped young wings. When the trembling young bird paused on the iron rung of a yard chair to rejoice and gather strength, I also sat down, feeling the need of rest.

❧ THERE IS some inconvenience in having a brown thrasher's nest under a sour-dock plant in your strawberry patch. You can't just bolt out carelessly and snatch enough berries for a quick supper. You have to walk carefully, so that if the mother bird is on the nest you will not frighten her off and can still walk within eighteen inches of her. She will sit there as still as the clump of dead grass she hopes you think she is, until you are close; and even then, if you continue quietly, she will look steadily back at you as long as her brown courage endures. Her brown body, striped with lines of brown and white dots, her long, wide brown tail, and the long bill make her easy to recognize.

When she is not on the nest (as she sometimes is not, for it takes both parents to make a living for the four bottomless mouths in the nest), she is not far away and is keeping watch.

She speaks a limited language, expressive and easy to understand. She herself told me about the nest. When I was getting too close to it, not knowing it was there, she was sitting on a gate behind me and she hissed. It was not a threat, like a cat's hiss, or a snake's, but merely a warning to me to be careful where I trod because the nest was on the ground and I might easily have stepped on it. Looking in that time, I saw the four fuzzy, bare-looking young birds. Their yellow-rimmed, square-cornered mouths were wide open, waiting for food to be dropped into them. The brown-thrasher mother has a special word for them. It is a snap and is onomatopoetic. It means "snap shut and stay shut."

What is impressive is that the young birds understand and obey without hesitation or argument. The purselike mouths snapped immediately shut. For ages the survival of young brown thrashers has depended on this instant comprehension and obedience.

BIRDS START on the cherries as soon as the fruit begins to show color. After two weeks of their happy foraging, the tree was still full of not very ripe cherries. Better half-ripe than none at all, I decided, and went out with a big kettle.

The birds retired in angry eloquence, expressing their opinion of cherry thieves, but I didn't care. I like to pick cherries. On a sunny day I like the cheerful sight of green leaves and shining red fruits in the pattern you see when you look upward, through the tree, into the sky. I also like the castlelike feeling of being up in a tree. It's probably something of the feeling children have when they build a tree house and go up into it, hiddenly.

Yesterday was a gray day. The plump-cheeked cherries were bright against the gray. Little tatters of wind shook water down on me, like a woman sprinkling clothes to iron. It was a wonderful place to be alone for a little while, thinking my private thoughts. Everybody has to have some occasional solitude into which he can retire to groom his defenses, think his thoughts, sort out his beliefs and strengthen them. It is not a question of whether he shall have a place of solitude, but of where the place shall be. It can be a game, a bird's song listened to, a song made, a woods to walk in, a field to till, a religion, a place of noise or a place of quiet. Or a cherry tree.

When I left, neither I nor the birds could miss the big kettleful of cherries I had taken.

I underestimated the birds.

When I went back the next day, the tree had been cleaned utterly. Not one cherry remained, either green, ripe, wormy, or bitten. It was a fine example of co-operative bird effort. But henceforth when somebody speaks of "eating like a bird," I want it explained what the bird is eating. A bird eating cherries eats like a glutton.

WE WERE GOING to have a picnic supper up in the north pasture beyond the woods for Alan and Kathy, city cousins from Maryland.

"It'll be a fine place; wild grape is in bloom now. I smelled it when I was up there yesterday stretching barbed-wire fence," said Dick.

On the low flat-bed wagon, tractor-pulled, we went up the hill pasture, over the terraces, toward the woods. The cows watched curiously as we passed them at the self-made pond on the hilltop where two dead trees standing in the water give a touch of Salvador Dali surrealism to the farm setting.

As soon as we entered the narrow lane at the woods edge we felt the coolness and smelled the wild-grape perfume. It hung on the air like smoke from a campfire. It is one of the most delightful and distinctive of all the farm's many wonderful fragrances. Neither as heavy as milkweed, nor as sentimental as sun-warmed roses, nor as emotional as lily of the valley, it is light, pervasive, with a heady bouquet. It drifts over wide distances, elusive and seductive.

You can travel a country road day after day not seeing the brown, coarse-barked, ragged vines sprawling from trees. And then suddenly some June morning when straw-berries are little and shiny-necked and vanishing, and the mock orange is in full snow with a ruby-throated hummingbird skiing over it, you will catch a whiff of that wonderful, insistent anouncement that wild grape is in bloom, and you will stop dead in your tracks and be glad you lived to that day.

Dick reached out and broke off a long piece of the blossoming vine, as we rode through the woods, and handed it to me. In the unspent late sunlight, the insignificant yellow bloom made an almost metallic glint against the pale green of the grapes, which had already begun to form in a cluster but were no larger than the head of a pin. Clusters of bloom, like fuzzy, small clouds, hung down under the grape leaves. The perfume is contained in the individual small, flat, yellow flowers, which are shaped like the centers of tiny daisies. They never last long; the fragrance is over within a few days.

We all took deep breaths as the wagon moved slowly along, barely having enough room on the narrow, tractor-made road. Honey locusts with new soft, red thorns and glossy, dark-green multiple leaves reached out toward us. Kathy felt very daring as she pulled walnut and honey-locust leaves from the trees. Alan broke off the edible new sprouts of sassafras to eat as we moved along. We went slowly past the sinkhole, slowly past the bee tree, slowly down the daisy-starred lane to the edge of the creek field.

There we built a campfire and cooked a picnic supper and stayed until dusk so that it would be dark as we drove back to the house.

NATURE has equipped the English sparrow with courage and persistence, self-discipline and protective coloring; and then, as if suddenly afraid that with all this the sparrow might outsurvive everything else, she endowed this drab, unpleasant-voiced bird with enough plain foolishness to give the rest of us a break.

All the time I was cleaning out the shed where the farm freezers, abandoned furniture, glass cans, magazines, and old toys are stored, the sparrow kept uttering, at intervals, a short, annoying shriek.

When this finally became intolerable, I had only to stop work and listen in order to locate the origin of the noise and, consequently, the nest. It was on a rafter, between two large sheets of cardboard. The untidy ends of sticks, protruding like shreds of coleslaw out of a sandwich, showed exactly where to look.

When I lifted the top cardboard, two young birds shot out, one flying straight ahead, one straight to the back of the shed. Their feathers were fully developed, but their self-discipline was not. The mother's was, however. She stayed there, head bent down as if dead, until the young birds were safely gone. Then she also shot forth. It must have taxed her self-control severely and required great courage to wait. But then, if she had kept her foolish voice still, I would never have suspected the presence of the nest in the first place.

IN THE AFTERNOON, having baked a batch of sour-cream cookies, I took some and a jug of ice water back to the north side of the farm, where Dick was harrowing corn, and in the adjoining field Ralph Lewis was mowing hay. Rose went along.

Beyond the barn I stopped to check on the many small crates and housings in which Joe has his summer's poultry interests. A crate on four tall stilts held four pheasants; they'll have to be turned loose soon because close confinement makes cannibals of these wild birds. The half-grown young mallard ducks honked cordially, thankful for more drinking water. The Bantams and their chicks ran out of Rose's way. Albert, the one remaining gosling of four hatched (a sow ate two and one just died), stood up and whistled. Why does a goose whistle? Albert is the farm's clown.

As we walked up the hill toward the field, I was pleased to note how the old, unused road is now healed of its deep wounds. From the top of the hill I looked back down at the house, which is all but hidden in the clustering of too-many trees. I wondered why old-time farmers built their houses at the foot of the hill where the water was, instead of on top, where the view was. The view here was wide and abundant, with wooded hills stretching far out into the other farms, the neighbors' white houses and barntops showing, and an infinite blueness of sky, in which a hint of rain was gathering now.

Near the house there is a cased-up well into which a vein of cold water runs ceaselessly. The builder of the house set it there for the water's sake, but in spring, and in heavy rains, the vein carries mud and fills up the well. He finally had to build a cistern, which is still our source of water for the house.

Rose was tired. We sat down to rest under an ash tree,

and I noticed how deep and lush the pastures now are. When I first saw that pasture, it was bearing a crop of wild brambles and desolation. In the heavy rains the fields mourned, the narrow creek in the front yard filled suddenly like a teacup under a faucet, and water running along the side ditches was brown with the sad knowledge that some field was migrating hurriedly. Water from this farm flooded the driveway of adjoining farms at a lower level.

Now the cattle can hardly keep the pasture cropped.

A faint continuous thunder came down from jet planes too high in the sky to be seen, and the harsher sound of real thunder sounded nearer, in the west. When I reached the cornfield, Ralph and Dick had stopped working and were trying to decide whether to mow more hay and risk getting it all wet before it cured, or to stop and bale what was already cured. Rose and I voted for waiting; in the meantime, we all ate cookies.

POOR RICHARD'S Modern Maxim: "Sit close to me," said Dick, as we drove home in the twilight, and added chuckling: "When they're young, they sit close together because they're in love. When they're old, they sit close together so they can hear each other."

IT IS ONLY at old-fashioned places in the country that people build a gnat smudge when they want to sit out on the porch or in the yard of evening. It is more modern to use insect-repelling candles or spray the lawn with chemicals to eliminate fleas, chiggers, mosquitoes, and gnats. You can hang chemically treated collars around the outside light bulbs or buy electric burners that clear the place temporarily of all insects. The world gets better all the time in ways of combatting insects, and the insects get steadily less destructible.

The smudge doesn't pretend to eliminate them. It merely creates authoritative boundaries for them.

It is a smoldering fire held down to a column of pungent smoke. It discourages mosquitoes and those extremely small black gnats whose bite is so much larger than their bodies.

The smudge is laid in a leaky old dishpan or kettle that has outlived its kitchen usefulness. First one puts in some dry chips that would like nothing better than to burst into flame and will therefore continue persistently in hope, despite the discouragement of wet grass or weeds piled on top of them. From this smoldering base smoke comes out and disinvites gnats. If the acridness of the smudge makes conversation and breathing difficult for the porch sitters, they resign themselves to the fact that silence is golden and the gnat smudge was not meant to be a conversation piece.

❦ IF YOU'RE in a hurry to get from here to the church, you can walk it in eight and a half minutes; but if you're trying to rid yourself of thoughts about atomic fall-out and strontium 90 and human sperm banks having to be established in case man's effort to annihilate his kind comes near to succeeding, then you can profitably take anywhere from half an hour to half a day for the walk.

The diversity and abundance of life contained in so small a distance is reassuring. Nature stores her capsules in small area and endows each with an unquenchable determination to survive and expand when the cue comes.

On the bank near the road a young tulip poplar had finished blooming and dropped its orange and green petals on the plants that pressed against its base. There were wildgrapevines there; a tiny redbud with pinkly young heart-shaped leaves; five-leafed ivy; and a long bittersweet that, lacking tendrils and thorns, had curled itself around the others for support.

I stopped to take note of harsh-foliaged whitetop, with daisylike flowers; milkweed in bud; honeybane with pink stalks bearing green flowers and a sweet fragrance that disappeared almost as soon as the flower was broken off. I saw Solomon's-seal with white bells hanging; tall, shaggy plantain going to seed; sour grass; dandelion; wild lettuce; tall, thrifty horseweed offering its long tassellike blooms for anybody inclined to go fishing with them for bait; goldenrod with no sign of flower bud, although insects had already stung the stems, causing the bulges that little girls use for stick-dolls' heads. The small wax-white flowers of pokeweed now disclose tiny, flat, green pumpkin-shaped pods within. Wild parsnip is tall, blooming in clustered, smooth, yellow flowerets somewhat like the flowery heads of Queen Anne's lace. I counted five kinds of clover; saw stiff, white yarrows with ferny foliage; sticktight burrs blooming now in asterlike flowers, pretty to see.

Raspberries are coloring; mulberries have fallen on the ground like furry fat caterpillars; imperfect green walnuts fell, giving the better ones room to enlarge. Curly dock is in bloom, its tiny flowers like drops of white paraffin dropped on green petals.

Greenbrier grew on thick, squarish stalks near a glossy ivy that will later develop a small blue berry. There were the grasses: Kentucky fescue; purple-brushed brome; timothy; bluegrass already gone to seed; orchard grass; wild barley with thorny heads; wild rye sparsely bearded like the hair on a sow's back; blue-green wild oats. New sassafras shoots had sprung up where already one growth had been cut this spring; hackberry, black locust, and Osage orange had reappeared. All these, pressed close together in small type, were only a tiny paragraph of what nature has written on the short road up to the church. And this was not all; there were thousands more. Nature has no intention of being exterminated, however gifted man may turn out to be in matters of destruction.

IT WOULD BE no wonder at all if a farm girl grew up to become a designer of exciting dress fabrics and accessories, all based on the commonplace things she met in her childhood on the farm.

The transparent hollyhock flower, with its heavy-corded, silken texture inspires a dress material to be trimmed with buttons that copy the immature seeds of the hollyhock. Strip off the fuzzy, pale-green covering of the seed pod, and a flat disk is disclosed, the size of a quarter and ringed around with unripe seeds giving a carved effect.

Buttons could also be modeled on the greenish-gold hard-shelled beetles; the spotted black and orange lady-bugs; the mosaic, carved backs of turtles; or almost any kind of seed or seed pod.

On a dress dark as a moonless night there should be buttons gleaming like sheep's eyes when car lights shine on them late at night in the lot: a greenish-blue, burning with wonderful opalescent fire.

Consider also a velvet suit the brown color of shelf fungus on old trees in a late-winter woods. At the hemline let there be sewed many rows of tiny velvety buttons the color of the scarlet spiders that live at the foot of the fungus-bearing tree. For a truly exciting dancing dress consider the rosiness of new, uncurling grape leaf. Velvet, certainly.

In her everyday chores, such as the preparing of vegetables and fruits for cooking or canning, a farm girl can find fascinating originals. Slicing green peppers for a cottage-cheese salad, she dreams up a dress of heavy, off-color white, with dark-green squarish rings and shadows in it, like the little flat, unripe pepperseeds. Canning corn, slicing off the creamy, shining grains in a clean slab, she thinks of shiny silk, white and yellow, the design of which duplicates the rows of not quite square tilelike cut grains. With this she will need a purse and high-heeled pumps of rough silk the color of new corn shucks.

For a winter dress, let her have a heavy, white, white silk, as softly white as new snow fallen quietly in the night and all crisscrossed with the telltale shadowy foot prints of little animals that walked abroad in the delicious night.

For a spring dress, a billowy, full-skirted soft green like newly leafed-out willows, and lace like the foam on a woods creek frothing itself against clean rocks.

If she wears blue jeans, or a feed-sack apron while she designs these creations, what does she care?

"THERE's a squirrel in that middle maple, on the snag 'way up in the trunk," Dick told me when I took up the milk bucket he had forgot. He leaned against the top slat across the barn doorway and pointed. "It's a young fox squirrel, about half grown."

I could see the squirrel's outline. He was sitting on his hind legs, eating maple seeds and dropping the emptied pods down on the ground.

The rain kept up a slow, running-water tapping on the metal roof of the red barn. "This is interesting," Dick went on. "I've often wondered what squirrels do when it rains."

When the squirrel had emptied his cheeks, he came down the tree trunk, front feet first, holding his proud, bushy tail high, and gathered up more seeds. Another squirrel came and joined him. For a short time they played together in the rain, like pups.

The scent of honeysuckle was sieved up through the wet air from the wildlife corner of the front yard. Out in the pasture, cattle swung their tails slowly, as if they were enjoying the feeling of rain stroking their new, clean hair.

"Squirrels raise two litters a year," said Dick, who never shoots them. "Those two down there playing are from the February litter. If you shoot a squirrel in January, you're probably wiping out a whole family."

MILKWEED and chiggers are both in bloom when raspberries are ripe, and you meet all three in the raspberry patch. All are prepared to defend themselves against humankind, but in different ways.

The raspberry defends its fruits with thorns which puncture and tear the skin but which can be thwarted if the picker wears sox, blue jeans, and long sleeves.

Chiggers are subtle and more deadly and return no good whatsoever for any favors. They strike relentlessly from their invisibility. Their reluctant host can outwit them by taking a quick bath immediately after getting out of the raspberry patch.

Milkweed has the most surprising and admirable defense of all the three. Its tall stalks are set with pairs of elongated leaves that have a rubbery surface. The flower buds are like popcorn balls, small and green at first, then taking on a rosiness as they enlarge in preparation for blooming. The five-petaled flowerets, each petal curved like a trough, are suspended from individual short stems in the popcorn ball. The fragrance of milkweed is communicative, much too heavy for long close association. I watched while its summons was answered within half a minute by two honey-bees; a sweat bee; a shield-shaped large beetle, and one a fourth as large as an appleseed; a housefly; a tiny black gnat; a red-orange tiger beetle; and two long, narrow-bodied flies with gauzy wings folded against their backs, seeming as restless as two new convertibles parked at a curb.

Presumably all these get some reward for helping to pollinate the milkweed. The raspberry picker, however, does the milkweed no particular service.

The milkweed's intention is to ripen its pale-green pod, which is made like two rough-sided canoes glued together at the rim into a container. Inside the pod is the milkweed fish, the scales of which are flat, brown seeds, each having its own silk-thread parachute by which to travel, wind-borne, to a new home.

The raspberry picker contemplates the milkweed's possibilities and finally breaks off a stalk to carry into the house for research. Any part of this plant, broken off, drips a milky, sticky juice containing latex. A farmwife picking raspberries while being attacked by thorns and chiggers can dream opulent dreams of the money-making uses of milkweed latex. Rubber for golf balls, tractor tires, overshoes, bathing caps; maybe she could at least produce her own girdles. But the milkweed has a plan for defending itself. On the back porch a bouquet of it, put into an old black apple-butter jar, looked decorative but quickly took possession of the room. "This milkweed smell is too much for a small room; it actually makes my eyes hurt," declared Dick. It was overpoweringly sweet. What except milkweed would have thought of using sweetness to wage offensive defense?

SOME CHURCHES call it Homecoming. At Maple Grove it is called Big June and comes the third Sunday of this month.

The usual Sunday morning service is followed by a carry-in dinner at noon. There is an afternoon program, with music and speech, and no evening service after. The essence of Big June is reunion between friends who come every Sunday, those who come less often, and those who come only once a year.

The church was organized in 1876. The small, one-room, white-frame building, square and steepleless and simple as a kitchen towel, was built then; and the congregation of the old North Liberty church came in. Soon afterward stone walls were built, enclosing the building and a cemetery within a square acre. The flat stones for these dry masonry walls were struck out of hillsides in the surrounding woods and hauled out in oxcarts. The walls average three feet high, go two and a half feet underground to be frostproof and ground-hogproof. A rod a day was the stonemason's limit.

In those days services were held only once a month, but there were Saturday night meetings with religious services and a business meeting, and religious services on Sunday morning and evening. June being a good month for neighbors to eat together outdoors, that meeting soon became an all-day affair, a social event that was greatly looked forward to. People came in buggies, wagons, surries, on horseback, on foot.

The church sits like a neat little white-clad grandmother on the flattened top of a green hill. The Maple Grove Road curves sharply there, enclosing the church as in a loving arm. Between the road and the church's stone wall, a big oak and an old hard maple make shade for cars, as they made shade for horses hitched there earlier.

East of the church, outside the wall, tall ash trees stand on the thick green sod of the church's flat grounds, providing a good place to set the long white-painted tables that are carried out of the woodshed and unfolded to hold the Big June dinner.

The dinner menu sounds like the index of a recipe book. The farm women of the Maple Grove community are good cooks. Each brings the foods on which she has built her special reputation, the foods by which she makes her harvest dinners remembered. Eating there, you start with crisp hot rolls, finish with strawberry shortcake and glasses of iced tea, punch, or coffee. In between you could have had fried chicken, baked ham, chicken and dumplings, beef and new potatoes, new peas, tomatoes, tossed salad, coleslaw, green beans, baked beans, butter beans, at least ten kinds of pies and cakes, and almost anything else you can think of to eat. You can come back for seconds or fourths, and there is still food to be repacked and taken home, or exchanged.

Beyond the tables is the view of gently sloping green farm field, dimpled valley, wooded hills going into it and rising gaily out of it again and continuing on into the dipping, blue horizon toward town, or down the road in the other direction, toward this farm. North of the church, across the road, lie the hilltop fields of this farm. For me, the level strip of road from the church to where the hill breaks and starts down is a place of particular appeal. I love it in daytime, on sunny or rainy days. I love it on moonlit nights, coming home from town. When I die, I am going to haunt that part of the road and give special blessings to all people passing along slowly enough to feel the beauty of the place.

Preparations for Big June actually begin with the Council meeting the Wednesday before, when the women gather to clean the church. They scrub the floor, the windows, the woodwork; put back the long, cotton window blinds that had been taken down from the tall, clear-glass windows and washed, starched, and ironed. They sponge off the narrow blue rug in the center aisle, on which every Sunday morning the deacons and elders walk up to the small, square Communion table in front of the pulpit.

The Council women have brought their own ladders, soap, dust mops, vacuum cleaners and brooms. They have a good time visiting while they wash the blue-flowered glass chandeliers, the glass on the framed picture of Christ praying in the garden of Gethsemane. They shake the dust out of the American and Christian flags that stand properly at right and left of the preacher. They close and dust the big old Bible on the pulpit, and put the minister's sermon notes, left there from last week, into the catch-all cabinet of the pulpit. They dust the blue-upholstered chairs and set them stiffly to right and left of the preacher's larger chair. They turn the happy side of the attendance banner out, to read: "I am early, what a pleasure!" (On the other side it says: "I am late, what a pity!" The young secretary is supposed to turn it when she fills in the attendance bulletin during Sunday school.) The Council women note that last Sunday's attendance is recorded as thirty-three, the record as a hundred and fifty; and everybody hopes the church will be full on Big June Sunday.

They clean off the long, gray-painted benches, and one woman, finally giving way to a long curiosity, crawls under a bench to see how much is cherry and how much is poplar. One woman stands on a ladder to wind and set the big wall clock. The blue-edged new songbooks are placed in the backs of the benches. Then the women leave their mops and brooms and buckets sitting around while they take time for dinner and after that for a short business meeting.

On the morning of Big June, they bring flowers from their gardens. At noon, while women put dinner on the tables, children run, play, shout; men talk farming; older people walk through the cemetery, not morbidly nor even sadly but because there are people there they loved and remember. The cemetery is well kept, pretty with flowers, and peaceful with grass and birds, stone walls and vines. There are small American flags on the graves of soldiers.

There are peonies, late iris, geraniums in bloom, and a sense of reunion all over the place.

"THE ROAN COW had a lovely calf yesterday morning," Dick told me as I was feeding the chickens.

"Oh, I'm so glad you didn't sell her to that woman last week," I exclaimed. Farmers have a superstition that if somebody wants to buy something and the farmer won't sell, some disaster will overtake the animal.

"I saw the calf yesterday evening," he want on, "but today when I was fixing fence up there I couldn't find it. And neither could the cow!"

As we walked up to the woods together, hunting, he went on with the story. The cow had bawled and searched. When he went to help her, she always led him to the walnut tree, but there was no calf there. "It might be sleeping someplace, and just not hungry enough to answer," he suggested, "but that's unusual. A calf stays where the mother leaves it."

We searched along the shady rim of the woods and in the shallow dry pond and in the rocky creek bed. No calf. The cow was grazing in the short grass of the next field. She knew at once what we were there for, and she appreciated it as the mother of any lost baby would. She really needed the calf; her large udder swung against her legs, its quarters extended like the legs of a milking stool. She ran up the slope toward the walnut tree, bawling harshly.

Passing the sinkhole there, her voice suddenly changed

to a note of deep tenderness. At the same time Dick exclaimed: "What's that noise?"

From the stony depths of the crevice came a faint answer. We looked down between the sides of stones and past the rolls of rusty woven wire fencing dumped there, and we saw a little red tail switching expressively.

Dick climbed in. It took a rope and a flashlight and the help of Ralph, who lives near the pasture, but we hauled the calf out, uninjured and hungry. Within two minutes the mother cow had smelled it all over, and it was feasting noisily, foam flying. We were quickly surrounded by a ring of curious and sympathetic cattle.

My own curiosity is unsatisfied. Did the calf fall into the sinkhole because Dick gently declined to sell the cow? If he had refused more emphatically, would the calf have died? It's this kind of uncertainty that keeps a modicum of superstition alive, but keeps that modicum under control.

THE BOLT BUCKET was within plain sight in the barn, and the wren building her nest there was working as hard as any mother who has a daytime job and does the laundry and housekeeping at night.

A wren will build in any kind of cavity, using sticks, grass, and stems to make a small hollow. Then she lines it with soft material, sometimes even her own brown feathers, to make a place for the six to eight brown-speckled eggs she intends to lay there.

"She just launched one brood last week," said Dick, "You'd think that would be enough for one summer."

The wren flew in as if she had a deadline to meet. She carried a wisp of straw, which she dropped into the nest; then she hurried out again.

"Stand up a little nearer," urged Dick, "and we can hold her off and get a good look. She sings with the grass in her mouth!"

When she returned, almost immediately, with more grass, she was not afraid but was annoyed because we were too curious. She perched on the rafter above the nest, and we had a good look at her pale-gray underbody, brown jacket, and the white markings near her long, flat, restless tail. The grass she carried had a long, forked seedy end, which gave it the appearance of an olive branch. With her mouth tight shut to hold the grass, she suddenly sang: "Chee, cheree, cheree, cheree!" For this we left her in privacy.

THE AMOUNT OF wildlife killed along country roads by cars is saddening.

By daylight, on the way into town, you see the citizens of the wild world and slow down to avoid striking them. It may be two rabbits conferring in the middle of the road at sundown; or the small, gray-brown least chipmunk (which farmers call ground squirrel) racing across to the weedier side. His long, flattened tail, held upright, helps a driver see him in time to avoid hitting him. It may be the brightly-mottled box turtle crossing as if he had eternity for it; or a fat, warty toad sitting there catching gnats faster than the human eye can follow. Everybody will slow down, not from pity alone, but dangerously if need be, to avoid colliding with a skunk on the road.

Seeing the wild creatures is a pleasure. "Isn't he cute?" you say. On the way home you may see his overtaken body lying in the road. Perhaps note of it has already been taken by the buzzard gracefully riding an air current above the road.

When Grace and I were going to town yesterday, I swung out to miss a ground hog that had not been missed by an earlier car. She said: "How sad! When Mirl and I came home last night, the car lights picked up the bright gleam of eyes from the middle of the road. Mirl slowed down, and we saw that a raccoon had been killed and was still lying there. Its mate had come out and was sitting beside it. It was the saddest sight; a raccoon has such a human-looking face, you know."

A BUZZARD was flying over the corner of the east field in a slow, descending glide as I came home from Ellettsville, and I stopped the car to admire him. A large bird, with dull feathers and ungainly bare head but strong supple wings, he was so graceful in flight that I would almost have been willing to exchange identities with him.

From great heights these valuable scavengers are able to see small, immediate morsels on the ground, for eating.

Farmers do not believe they locate this food by smell. "If you hit a snake with a mowing machine, a turkey buzzard'll find him right away, even before he has time to develop an odor," said Dick. "I think they have telescopic sight, like eagles or owls."

Perhaps the buzzard realized that I was watching him. He ascended again. He accomplished this by a very few deeply curved pumpings of his wings. It looked as effortless as flicking the ashes off the end of a cigarette.

As I watched the bird make the few, necessary strokes, I could actually feel the lifting sensation. From my armpits to the tingling heels of my palms, I could actually feel that wonderful sense of becoming air-borne and ascending higher. For that brief moment I had the thrilled sensation of flying and of freedom. A beautiful, memorable gift from a bird whose name is often used as an unflattering epithet.

As QUIETLY as possible, so the family would not waken, I came down the creaky, crooked stairway.

I often walk at night.

On a moonless night, feeling too much enclosed, or on a night of intense, full-burning moonlight, because its beauty is too much to endure quietly in the house, I walk. The wish to walk outside in the yard, or up to the gatepost in the barn-lot, or on the road past the mailbox, is like wanting a drink of cold water.

In the comprehending way of a dog, Rose walks with

me. She got up silently from her sleep on the iris bed by the east wall and followed as I crossed the yard. When I paused to listen to the drip of dew from the maple leaves, Rose pressed the cool moistness of her black nose against my hand.

The driveway was immersed in deep shadow, but the road gleamed in the moonlight and the mailbox gleamed, too, as we passed it. The charm of night-walking is not that it solves problems, nor that it yields up great truths, but that it offers a widened, less acutely personal view of the problems and suggests that somewhere in the shadows the great truth is waiting, almost within the mind's grasp.

Still restless I turned and walked back across the yard and up to the barnlot. When I climbed up to sit on the big gatepost, Rose stood waiting, patiently.

A farm at night is a different world than by day. It seems thoughtful and ancient. Perhaps it is the moon's light, ancient itself, that gives that sense of timelessness to familiar things at night, to the cattle lying down out in the pasture, to the dew-drenched trees, even to the house, quiet and shadow-cradled at the foot of the hill.

Sometimes on these expeditions I feel I have gained a faint sense of the destiny of man. His mind can perhaps be likened to land; his present cultivated intelligence, compared to his potential, may be like a tiny cleared corn patch in a great, unexplored pioneer wilderness. When all the forest is explored and brought under cultivation—which will use up all the time of eternity, no doubt, and is intended to—what will man then be? Will he not be truly free? How long is eternity? We cannot even conceive of a time or area or freedom that has no boundary.

Rose's eyes, looking up at me, were amber in the moonlight. I jumped down from the gatepost and stroked her gently on the underside of her throat.

What do you think, Rose?

She thought it was pleasant to be stroked. She thought it would be pleasant to go back and sleep on the iris bed. She thought it was time for me to go back to bed upstairs. It was. We parted at the back step. Good night, Rose-in-all-the-world, night-walking could be lonely without you.

THE HOGS here are always Durocs, a breed that is red and good-tempered. The boar is always named Tonto. There are now five Mrs. Tontos in the barn, and each has a family of little copper-colored Tontettes.

Last week I mulched the raspberries with manure from the barn. One afternoon while I was working there one of the Mesdames Tonto put her head out of the barn window and greeted me with an amiable: "How?"

We conversed for a while then. Mrs. Tonto would have liked me to scratch her head, but I was too busy. I tossed her a broken piece of corn I picked up from the ground. This endeared me to her, and she remembered it the next afternoon when I went back.

We talked again, exchanging news and views but no recipes, particularly no ham recipes. I observed her skin, which is coarse and dirty. Her hair is dull, sparse, straight. Her ears flop down, nearly covering her eyes. Suddenly curious to know what color eyes she has to go with that skin and hair, I bent down and peered intently below her ears. She bore my scrutiny with friendliness and dignity and looked back at me. Her eyes are a reddish brown, with pale, scanty eyelashes, perfect for her coloring. Afterward she did not resume the conversation, but retired with a sow's genuine poise and dignity to her parlor in the barn.

JULY

[JULY]

JULY comes with an armload of abundances like a farmer coming home from town on Saturday afternoon.

A mellow, generous, middle-aged, productive month with a thought for the past in its misty blue mornings and a thought for the future in the first tentative creakings of cricket and jarfly—July has a thorough, gusty enjoyment of the present moment.

Now black raspberries are diminishing, but red ones are taking their place, and blackberries will soon follow. Already the first Yellow Transparent apples can be picked, to unburden the tree, and made into pale-green applesauce. The garden, lavish with green beans, hints of roasting ears, cantaloupe, cucumbers to be sliced thin as paper and crisp as a new five-dollar bill. The smell of new beets cooking in the kitchen fills the farmhouse. The first ripe tomato this week was an event, but two weeks from now the planters will "eat them as common things," as the Book of Jeremiah says, describing abundance.

In July there are no fires to keep up any more. You can use the whole farmhouse freely, the whole yard, the whole farm, the whole day. Freely, freely, freely. Freedom is the essence of July; independence is its rising up and its lying down. I love July. If I have to make some criticism of this too-short month (it has only thirty-one days), all I can complain of is its sogginess. In July you finally have to make up your mind whether you prefer the desk and bureau drawers stuck shut or stuck open.

POOR RICHARD'S Modern Maxim: "That's a farmer's job," said Dick when somebody complimented him on his acute observation, "to see everything. If he don't see it, he's done."

IN JULY the wheat is combined; the hay mowed, raked, and baled; the corn spray-weeded or cultivated and laid by. On the farm nothing is idle, neither root nor corpuscle. Meals are late because the men have a last load of baled hay to get into the loft before the storm breaks out of a tumultuous dark-blue, Rip-Van-Winkle-thundering sky. Or they are early, so the balers can start to another farm two miles away for a custom-baling job. (Silo fillers get dinner served, but, except at the homes of special friends, custom balers and grain combiners provide their own dinners.)

In July there are so many things to do you have no time to worry about any of them. There are so many abundances to harvest and richnesses to see that you can leave the jelly-making and go without compunction when someone rushes in to ask: "Have you noticed those red lilies in bloom down at the fork of the road?" Or: "There's a deer crossing the springlot field. Or, coaxing: "Come up and watch me feed the skunk; she's so tame she doesn't throw scent any more." One abundance is as good as another, or better.

July mornings have a sense of timelessness that is worth getting up to see. It is in the coolness of the early, misty, pale-blue and gray mornings; in the wide reaches of green, wooded hills, ripening wheat fields, and restless blue-green cornfields. It is in the beautiful green hayfields, when the hay has just been cut but has not yet been raked up into windrows.

July has a sense of history. It probably was not by mere coincidence that in various years the sharp and bitter clashes came in July, and that it was in July that America first asserted its independence and determination to be free. In July the thought of giving up independence is intolerable.

THIS IS the day on which, in America, freedom rings. This is the fourth of July.

Early this morning Dick and I were sitting on the front porch enjoying the supersaturation. The air was so full of moisture that if one spoke sharply it was likely to start a rain. (That's one definition of supersaturation; another is three 4-H events in one week.)

We sat on the porch, both of us aware of our many blessings but not mentioning any particular one by name.

In this idle and wholly profitable way of spending time, with no words cluttering up the beautiful silence, I suddenly had the conviction that if I ran out quickly, without stopping to wonder, I could run straight up the side of the big maple as I had just watched a chipmunk do. There I could sit on the lowest limb and swing my feet and enjoy the astonished look on Dick's face. I would have to run fast, so that the ascent would be an accomplished fact before rationalization set in. After I had got up there, I could explain it by scientific laws, gravity, inertia, horsepower, and so on; but at the moment of getting there it would be simply because I had followed the conviction without stopping to question it.

FARMERS CELEBRATE the fourth of July by working right ahead in their independent way. In the evening, well before sunset, the farmers of this community go to town, where every year at the university stadium the American Legion provides a wonderful, free display of fireworks.

The Legion gets the money for this from a carnival but at the stadium gates also provides a large kettle into which you may toss a freewill offering, if you wish. The crowd comes in a mood of gaiety and patriotism.

The wide ovals of bleachers surrounding the green playing field fills steadily. There is a constant seething back and forth, up and down, as people come to get seats or go down to the refreshment stand to buy soft drinks. Parents come, bringing babies in arms or small children. Teen-agers come with dates, hoping to sit where their parents won't be. Older people come. The many foreign students taking summer courses at the University also come; this typically American gathering must be almost as fascinating to them as the fireworks are.

As we sit there, waiting for dusk to gather like cream on a crock of milk so the fireworks can begin, the bleachers fill. The clothing is of so many different colors that the effect is of a great, oval mosaic of color. There is some predusk entertainment usually: a color guard, the champion drill team; baton twirlers, a phonograph playing popular music over a loud-speaker.

Then the Star-Spangled Banner. No matter where, nor how often, I hear our national anthem, it evokes a combination of tears, goose flesh, and prayer from me. "The Lord did set His love upon you, because the Lord loved you and redeemed you out of the house of bondsmen."

The first soft burst of fireworks is like the snapping of a giant rubber band. It is experimental—is it dark enough yet? Yes, it is. The first shot bursts above the bleachers in a great bronze dahlia of light; blue stars fall; the crowd emits an ecstatic "ooooooo!"

The popping continues and increases. Streams of light

alternate with violent sound. Small children shiver deliciously and press against the arms and shoulders of parents. Babies cry, not yet realizing that freedom is a violent blessing and independence an explosive brilliance.

There are large displays of light: battleships exchanging shots; fountains streaming; Niagara Falls pouring. The faces of Lincoln and Washington appear in light. We see the American Legion emblem; Old Glory; and, in appreciation of the University's courtesy, an IU, as nearly as possible, in cream and crimson light. There is always the touch of humor loved by the crowd: the model-T whirligig that won't start, that fizzles and sputters and finally does start with a triumphant intensity; the fish swimming confidently across Niagara Falls. When the fireworks are spent, at last, the announcer's voice comes out of the darkness, urging people to drive safely on the way home; and the crowd leaves, full of renewed independence and content.

"WHAT I called you for," said Monta, whom everybody calls Montie, "was to ask if you have green beans to use. If you haven't, I'll send up some with Iry as he goes to work in the morning. I've canned eighty quarts, and some for Odette besides. Got any cabbage? I'll send a head of cabbage, too. It's jist down here, going to waste."

Montie is instinctively generous and loving, also forthright and practical as a broom handle. She likes bright colors, lots of perfume, long earrings and television. She is dependable; if the RFD club plans to have a bake sale, Montie will not forget to bring what she promised, and something more besides. She makes the best chicken and dumplings in the world.

"When I'm dead," says Sairy Brown, who lives in the 132-year-old stone house down the road, "I just want you to pour a kettle of Montie's stewed chicken and dumplings over me and leave me alone to enjoy myself."

Montie laughs often and freely and intersperses her laughter with philosophy, sympathy, news, even flashes of anger, so that a visit with her, even if only by telephone, is a refreshment to my spirit.

That day she had cooked dinner and supper for a haying crew. "We had to eat in shifts," she said. "Some were done, while others were still picking up the last load. I'm as tired as if I'd worked."

From this we went on to a discussion of how farming has changed. Farmers like to talk about how much simpler and easier it is now.

"We still have one load of bales on the wagon," Montie went on, "but I was telling Sharon haying ain't as hard as it used to be when I was her age. We used to pitch it up from the ground with pitchforks onto the wagons, or haul it in, in shocks. We pulled it up into the mow with a horse, and it was my job to sit on that big old fat, sweaty horse and pull the hay rope out. I tell 'em they don't know what hard work is now." She laughed.

"When we got a hay loader, we thought it was wonderful," I agreed.

"It's easier now, but yet people don't seem to have any

more time than they used to have," concluded Montie. "Well, come down, Rachel."

The next morning Ira stopped with a peck of green beans and three heads of cabbage.

Dear Lord, let farming continue to improve forever, but let the neighbors never change.

DO TOWN PEOPLE have any nuisance equivalent to the oats bug in the country? The size of tiny black splinters, oats bugs neither bite nor sting, but make themselves obnoxious by their numerousness and by lighting on your skin as soon as you step outdoors now. Weightless as they are, you feel them unbearably on your skin, walking. Just walking, walking.

SOMETIMES in the still mornings of these wonderful sunlit days, when the very air is the color of thoughtfulness, the earth and the fullness thereof seem to contain some poignant message, ancient and persistent. Before daybreak, hearing the dove sing close to the farmhouse wall, her voice like a pointed finger, velvet-gloved but inquiring; or the whippoorwill, who is really only a mockingbird trying his hand at being a whippoorwill; or a sparrow repeating her thoughts in quick, swept syllables like a woman sweeping the crumbs from a tablecloth into her other, cupped hand; or in midmorning hearing the thrush singing like a stone being shaken in a big gourd, I hear the message, but I cannot quite interpret it.

First settler

I hear it when the limbs of the maple tree stroke the walls of the house, and the wind rises, and the soft brush of rain begins.

I see it in the sun-struck glint of cornfields; in the tiny restless lanterns of fireflies (less poetically known to farm children as lightning bugs) rising from the grass on July evenings, going on and off and rising higher each time until they reach the level of tall trees.

It is in the cut wheat field, brown and tidy as a hair-brush, and in the sharp, clear cry of the quail nearby. It is in the ragged blue patches on tall vervain stalks and the intensely bright blue rounds of chickory; in wild dark-blue larkspur; and in the dusky brown centers of black-eyed Susans.

I smell it in the elderberry fragrance drifting in from the pale ivory-colored bloom; and in the first velvety, ripe peaches; and in the first big drops of rain on the ground after long drouth; and in the sweetness of hay, curing in the hot windrow.

When I tried to express this to Mr. Hanger, bookshop owner and preacher, I could not. He nodded—and we both knew he understood.

As IF he had been taught never to turn his back on an audience, the snapping turtle kept his small, snake's head between us and his body, meantime backing off determinedly toward the concealing grass and weeds beyond the road.

Neither were we able to turn him over, though we managed to tilt him enough to see that the flat, yellow underside was brighter than the dingy brown carapace. He was eight inches across, but seemed larger because he was so hostile. When he opened his angry mouth to bite the end of Dick's shoe, we saw that he had no teeth, but his jaws snapped shut with an ominous click.

The long, pointed tail, the claw-tipped feet, and the small head rose out of voluminous, disheveled folds of skin that seemed several sizes too large for him. What looked like a dark, wet snail out of its shell, turned out to be an old healed wound on the turtle's carapace.

Snapping turtles are equally at home on land or in water, preferably muddy water. "I've seen as many as six or eight at a time, sunning themselves on a log," said Mirl Lundy, "but they dive into the water as soon as they see a person."

This one, having nothing to dive into, had to face us as he retreated. It is not really fear of the turtle that sends that peculiar shiver over the spine of the person who looks into those little eyes, bright with anger, or sees the triangular, snake's mouth open hostilely. It is the realization that this incordial, slow-moving, unbeautiful creature was here first —that compared to the 200,000,000-year existence of turtles, man is a newcomer, a mere squatter on the beloved earth to which the turtle has, perhaps, first and greater right. The shiver arises from the speculation whether the newcomer possesses the older citizen's talent for staying around.

Cornfield in the Rain

THE BOBWHITE sings from late spring on, but his voice has always the sound of summer in it. One hears in it the golden-brown look of ripening wheat fields, brushed softly by wind and drying in the sun. From a short distance the quail seems not to have to test the pitch before uttering his sudden, liquid, pure melody. But if you are close enough, say the distance of half the corncrib's length, you hear the soft, experimental, almost whispered first syllable, like a French-horn player making sure of his tone.

Quails have an extensive vocabulary. They sing loudly, "Bob, Bob White," and it is not the same at all as the loud "Bob White, White." These are songs of the male. The female sings chiefly a gay "tooie whee!"

They sing at evening to round up the covey for the night; and members of one covey do not, uninvited, enter to spend the night with a strange covey.

They sing also to establish their boundaries. On the garden gatepost this morning a quail was singing. He was answered every time by another quail north of the barn and one farther south of it. A line drawn between the three would have made a wide triangle, marking what is, unquestionably, their specific domain. The boundary fences of their domain are merely the sentinel voices, repeated at intervals all day, like a farmer keeping up his line fence.

FROM THE kitchen window, the cornfield at the hilltop made me think of a sea. The dark-green blades, wet from last night's rain, waved like rippling water. The height of the stalks varied just enough to make green peaks in the contour. If the sun had come out, the corn would have twinkled like water under the sun.

It was going to rain again soon, but I had a necessary chore in the field, so I crossed quickly through the sloping pasture toward the hilltop. The cows stared at me curiously. At Rose they stared less with curiosity than challenge.

By the time we reached the cluster of walnut trees near the pond, rain was falling again. It struck the surface of the pond, piercing it like a nail driven through a tin lid. The brown, freshly tilled soil endowed my boots with mud.

In the cornfield, the rain began like a small sigh. Its whisper was magnified many times as rain increased in speed and volume and struck the long, flexible, warped corn blades. It was a satisfying sound, deep, increasing; an ancient, far-off sound, mighty and portentous. The sharp-edged blades received the gift of rain, and bowed ceremoniously in acknowledgement, thus transferring the rain to the right place, at the feet of the ever-thirsty, paler-green stalks.

I walked through the curved field, enjoying the taffeta sound of blades brushing my sleeve. At the second clump of trees I sat down on a lichened gray stone. Only a sifting of rain came through the sturdy leaf canopy. The green walnuts, two thirds as big as a lemon, hung in pairs, in threes, and singly.

Fearful, as always, at the sound of thunder, Rose pressed close to my feet. We could look through the glass-thread curtain of falling rain and see the sloping field, the top of the barn, the red roof, and the galvanized tool-shed roof and the silo. The big trees in the yard had swallowed all but a corncob's length of the house roof in one great green gulp.

In the cornfield, being on a much higher level than the house roof, I could see far out into the misty blue circle of the horizon, into the first valley, now filled with gray fog. Beyond that rose another line of green woods, thinned by distance and rain to a corn blade. The next horizon repeated the step, and on and on, until finally a stairway led up into the sky, out of sight, and on into forever.

IT'S A good summer for cabbage. The heads are tender, succulent, fast-grown, and just now beginning to split from rain. It's time to make sauerkraut. ("Never make kraut in the sign of the lady with the flower in her hand," my neighbor Rena Dutton always says. "It will turn slick.")

When Catherine and Fanny stopped here this morning, Catherine had just finished making kraut. She cut it on the regulation kraut cutter. (Some like it chopped; Catherine likes it shredded.) The kraut cutter is a long wooden board with a sharp metal blade inserted at an angle in it, adjustable by a thumbscrew, to make wider or thinner shreds. For five pounds of cabbage, Catherine uses three and a half tablespoons of salt.

"Work the salt and cabbage together with your hands," she said, "until the juice begins to come. Then pack it into glass jars. Pour in all the juice, and screw the lids down as tight as you can."

"No matter, the kraut will begin to work and some of the juice will spew out under the lid, fur's that's concerned," contributed Fanny.

After the kraut is cured, it can be put into the freezer, if you have room. Old-timers remember how good the cold top layer was in deep winter when you went out to the summer kitchen to fill a bowl from one of the five-gallon stone jars. Beside it was the stone jar of mustard pickles, grape-leaf-covered, with an old white-china plate on top, and a clean stone weighting it all down.

ON THE back-porch wall, above the kitchen door, a symbol hangs. Whether a symbol of fact to be understood in the impartial scientific years ahead, or a symbol of the superstitious errors of farmers now, it is an innocent-looking thing, a small forked peach limb with which, fifteen years ago, Emerson Dutton witched the place for a well in the yard here.

Emerson was a practical, thrifty, Dutch farmer, a generous neighbor, respected and not overimaginative. As a water witch he was held in great esteem.

People who discredit water-witching refer to the forked limb, which may be peach, willow, or any pliant twig, as a divining rod and put it in a class with the little bag of asafetida worn around the neck to ward off illness, the ax under the bed to stop night sweats, the stepped-on, rusty nail greased and laid in the window three days to foretell development of lockjaw. No local farmer would think of digging a well without having the spot witched, whether he believes in it or not. Mr. Empson, local well-driller, consults the water witch first.

On this farm a vein of cold hard water rises in some spot in the hillside above the house. When Benny built the house, he dug a well in the yard and cased it up with stone to a depth of five feet, and every year cleaned out the silt that washed in, threatening to fill up the well. Emerson traced the vein from this well across the yard, across the driveway, and down the slope to a muddy spot near the soft maple, where a shallow spring flows the year around and also has to be cleaned of silt occasionally.

He cut the little twig from the old bent-over peach tree by the back porch, grasped it with both thumbs pointing up, the center prong also pointing up. When the prong turned down, he said, he was on the vein. As he stepped back, or forward, the twig wavered and the prong turned upward again. Later we dug where he indicated, and water was there; but stone was there also. We abandoned the digging.

People who regard this as pure superstition smile, and put the twig back up above the door, and turn to express amazement as the neighbor's six-year-old boy comes through the yard, driving the tractor. The boy will start school this fall and already knows how and when to shift gears—which he has to do standing up because his legs are too short for his feet to reach the clutch pedal when he is sitting. He recognizes bearings, valves, gears, and knows their functions.

He understands the function of the battery, though he does not understand its electric soul. Even in the mechanical age, there comes a place where understanding stops.

The imagination of every generation strives toward a more perfect machine that can be operated with less imagination. This, then, becomes the ultimate symbol—the small, illiterate boy on a powerful machine, knowing exactly where he is going and at what speed, while around him, invisible, deep and strong under the earth, flows the hidden vein, unwitched, pure, cold, awaiting the forked imaginative pliant twig.

MIDSUMMER is the heyday of insect life. Now sits the *Insect Field Guide* on the back porch table and beside it the magnifying glass that just fits the top of a small-sized Instant Coffee can, the better for catching and examining these small neighbors. There is abundance of fly, wasp, spider, ant, beetle, and butterfly to study. The examination of any one of them leads to increased respect for the whole animal kingdom.

Long after midnight a beetle interrupted my reading by hurling himself into my lap. He was less than an inch long, black, shiny as a new car, with dents in his covering. There were mottles of deep orange on his wings and head. The outer ends of his antennae were hung with tiny pale-brown cubes. I didn't want to take time to put him outdoors but saw no reason for killing him, either, so I picked up a small brass bell from the lamp table and covered him on the floor.

Presently, it seemed to me the bell moved. Then I was sure it moved. It went forward an inch, slowly, as if by great effort, then stopped. A few seconds later an escaped beetle made a dash for a hiding place under a newspaper, and simultaneously the air was burdened with an unpleasant odor, as if several people had sneezed all at once without benefit of handkerchiefs. My visitor had been a burrowing beetle, whose strength, persistence, and ingenuity enable him to burrow sufficiently that he can bury a dead mouse by burrowing under it. His defense, like a skunk's, is his ability to throw off an unpleasant mist.

BOB TELFER likes red raspberries for breakfast every morning and green beans for dinner every day in summer, but has only contempt for rolls. "Even wonderful ones like yours," Bob's wife, Wolcott, told me.

"How can you eat that gummy, soft stuff?" he demands, and, snatching a match out of some pocket, begins the process of lighting his pipe, which requires several matches.

Bob looks and speaks more like an Englishman than an Indiana farmer. He is a graduate of Princeton University, reads avidly, has aggressive political and social opinions. He operates a little bookbindery for quietness and a large farm for unquietness. His farm is one of the best and most beautiful in the community. The big old brick house, furnished in perfect taste, is constantly the scene of social gatherings attended more by town and university people than by farmers.

Bob's father was a Methodist minister and, whenever he can get a group of neighbors together, sooner or later he gets them to singing *Marching to Zion*, which is his favorite hymn.

He underlines his conversations by snatching his pipe out of his mouth and holding it by the bowl, using the stem as if drawing heavy lines under the words.

He likes fried potatoes. "Not that pseudo stuff made from cold potatoes boiled with the skins on," he insists. "You must slice them fresh, start them in a sizzling hot iron skillet with fat. When they're almost done, add a little water, cover tightly, and finish. They should have a crisp brown crust next to the skillet and be light, mealy, and hot on top of the crust."

Then, having made a profound analysis of the complete reorganization of county government, he described "the perfect summertime meal."

"I like green beans," he said succinctly, using his pipe stem to underline those two words, "picked fresh from the garden and cooked with fat meat until they glisten. They're better reheated the second or third day; taste beanier. Fill the plate with beans. On top of that slice a layer of fresh, ripe, garden-grown tomatoes; next slice a layer of onions and pour a little vinegar over them. For a side dish have corn bread and buttermilk. Start eating at noon, and eat right on all afternoon."

Bob is tall and thin.

Consider the lily

THESE ARE the lilies, these tawny-orange day-lilies one sees abundantly along country roadsides, in farmhouse yards, and in abandoned fields. These are the descendants, no doubt, of those to which Jesus called attention when He walked along the lovely green hillsides and narrow roads around Nazareth. These old-fashioned flowers of East Asian origin, which toil not nor spin, are old friends of farm women who like flowers that come up year after year at the appointed time, asking no favors. They are tough and pliant and tall. Their large, six-petaled flowers curl back like tongues of flame. They are not easily discouraged. One summer I took up a clump of them to transplant but being interrupted set them down on top of the ground and never got around to planting them. The roots reached down through the surface anyway, the lilies established themselves and have thrived thereafter.

And this is Jimson, another dear old friend of mine. It grows on tall, forked, purplish stalks in places where nobody has time or ambition to chop it out. It used to be a garden flower but proved too easy to spread and became shunned. Its leaves are shaped somewhat like large holly leaves. Its flower is an exquisite long gored skirt, deeply blue at the narrow beginning, becoming white and pointed at the outer-flaring edge of the gores. It has a pleasant fragrance. At the cooling evening of a blistering July day, as one walks down from the barn along a dusty path, the fragrance of Jimson immortalizes the moment.

July

NOBODY needs to pity the farmwife, however scratched and bedraggled she may look, coming out of the blackberry patch. She has got something more than blackberries from the time out there. There is no better place than the blackberry patch for that hour of solitude every farm woman craves now and then.

She has to guard herself against chiggers. Some people rely on close-fitting arm and leg coverings; some use kerosene-soaked string belts; some dust themselves with powdered sulphur. Some pick and hurry home to the bathtub.

Having thwarted the chiggers, she has mosquitoes to deal with. Mosquitoes can't bear brilliant sunlight; they prefer the cool shady places, where the best berries grow.

The berrypicker may meet a friendly snake. Being intelligent, she will not venture into a pasture where there are cattle unless she knows the cattle.

With all these hazards taken care of, she is ready to enjoy her vacation from housework. She takes a big bucket and a little one. The little one hangs from a belt around her waist, so she can pick with both hands. She need not fear the large, iridescent blue-green June bugs that hang drunkenly from overripe berries and roar with the tumult of bumblebees. In a good blackberry patch there are enough berries for her and them also.

I like picking blackberries. I enjoy the sights and sounds, the feeling of freedom and isolation it affords. I like the luxury of this delicious wild fruit to freeze, bake into pies and cobblers, and make into jelly and jam.

In the pasture across from the church, where I pick

blackberries, the brambles huddle in dense patches. The grass next to them is tender and palatable, because Dick mowed the pasture late last fall. The cattle like to graze right up to the brambles, and so they tread down the tall weeds, establishing fine, open paths for me to walk in. There blackberries are so abundant that it is not necessary to fight my way into the thorny heart of the thicket, though it taxes the thrifty spirit to go away and leave the dark, shining, juice-filled abundance hanging.

Wild blackberries, in any one patch, are as varied as people, and I don't know what explains it, surely not the soil, nor water or sunlight, which are the same for all.

There is the dark, big-seeded blackberry that clings tenaciously to its stem even when it is so ripe it squashes in the fingers. It is a waste of patience to fool with these. There are the large, juicy beauties that fall eagerly into the under-held, cupped hand and can be picked five at a time. They have superb flavor, vigorous, sweet, wild-tanged. There are the small-seeded, dry, firm berries with the perfumy taste. Prissy, smug, like people who never encounter either danger or disappointment or soul-satisfying triumphs. Not much good for pies or jelly, or anything else.

There is another large berry that is sweet and tempting even when so firm it seems barely ripe. These berries stand well apart on the stem and are extremely bright. You can know them by their shine; they look as if illuminated for a photograph in the foods pages. There are the small, juicy berries with the flavor and fragrance of gin, which grow preferably in full, hot sun. There are the berries with large white cores and few juice-filled knobs. There are short ones. There are beauties with almost no flavor. An experienced picker meets them all.

Hardly anyone under thirty picks berries any more, and I would not enjoy it as a bread and butter job. As a jam job, it has its rewards. Since no one else in the family yearns to pick blackberries, it is one place I can count on for that solitude that every loving housewife needs once in a while. It is not infallible.

One year when the children were little and Dick took them with him on the tractor, I decided to take an hour's vacation in the blackberry patch. I would be blissfully alone; nobody could possibly find me there to ask a single question, I thought, luxuriously dropping a handful of berries into the shining new bucket.

Almost immediately: "How're you making out?" Dick asked from the fence at the edge of the woods. He smiled lovingly, his face beaming like a full sun, and on either side of him, like bright moons, beamed two small, beloved faces.

"How did you know where to look for me?" I asked.

"Why," explained Dick, pleased with himself and them, "we saw the sun shining on your bucket." The vacation was over and not one single "having wonderful time, wish you were here" card had I sent to anybody.

THE CHIEF BUSINESS in life in the animal world is to stay alive, even if one has to preserve life by appearing already to have lost it. This is a toad's way, but I didn't know it until yesterday.

He was no longer than my thumbnail, and only two thirds as wide, but his instinct of preservation was fully developed. I saw him in the fencerow of the hill pasture, watched him making short hops. He watched me out of bright eyes set in a spotted drab body. When I moved my hand, intending to cup it above him without touching him, he instantly flipped himself upside down in a perfect imitation of deadness. From the first flip the little body landed white underside up, shrunken and still, the tiny mouth opened, the little eyes became suddenly dull and vacant-looking. As soon as I removed the danger, he flipped back on his going side and came to life again.

It is not always so easy for a small creature to turn himself upside down and right himself again. Yesterday I examined a stag beetle, a wonderful two-inch insect that looks as if carved from gleaming rich mahogany. Lying on his back, he was unable to get himself turned over by the use of his legs alone. He had to use his antennae. By stepping on one antenna and pushing with the foot on the other side of his body, he succeeded in righting himself. He did it many times. To a human, unable to pull himself up by his own bootstraps, the stag beetle's way seems worth taking note of.

CAROL, whose chore for the 4-H meeting that day was to demonstrate making biscuits, was dreading it audibly.

"Don't worry," comforted Grace, who taught her to play the piano, "they're all your friends; they know you and wish you well."

"Yes," said the girl quickly, "but I'd like it lots better if they didn't know me so well."

Later in the afternoon she came home radiant. "You dread it until you get started," she exclaimed, scattering her demonstration tools all over the kitchen, "and then it's not so bad. And when it's all over, it's simply wonderful!"

IT WAS NIGHT; everyone else was asleep. The farmhouse had been overtaken by what seemed like a deep, universal peace.

A beetle with jagged antennae and an oblong inch of black body came hurrying across the bare floor toward the small pine cabinet where we keep papers.

The beetle, Prionus, and no friend of farmers because he attacks the roots of fruit trees and grapevines, had a definite destination. He went under the base of the pine cabinet, and there ran afoul of spiderweb. Housekeepers detest the spiders that make this kind of web. They are small spiders with overstuffed round stomachs and a persistent faculty for quickly rebuilding what the baffled broom wipes away.

The web was only flimsy fabric, but it stopped the beetle and set him churning ineffectually like the wheels of a car in soft snow. He tried to get free by backing and turning his forepart as a car has to, and backed into a green raveling on the floor; it stuck to him.

Now a small spider darted down from the underside of the cabinet and stung the beetle. I jerked back on the raveling, pulling the beetle to safety, and let him dangle from the raveling while I examined him. On the floor again, he worked loose and immediately resumed his rush toward danger.

Under the web-strung base of the cabinet, he again became hopelessly entangled. This time the spider came down more quickly, struck, dashed back up to his hiding place and returned again. The beetle, bewildered and awkward, raised his forelegs to defend himself, but another, much smaller, spider came up from the lower edge of the web, gripped one of the beetle's long, plated antennae, and held it, while the first spider returned at quick intervals, savagely.

You are not God, I reminded myself sternly; this is life. Let this conflict alone.

For a few seconds the battle raged one-sidedly; then, without planning to, I suddenly reached out a long envelope and scraped the beetle back to safety and smashed the two spiders.

In his frantic effort to clear himself of web, the beetle beat violently against the paper, buzzing in a loud whine. When he was clear of web, he backed, reversing his direction as a car would, and started again toward the danger-strung cabinet base. I let him go. I knew there would be other spiders and other webs on the other side of the cabinet, only eighteen inches from where a miracle had already rescued him. But if other spiders killed him, it would be his destiny. What else would have impelled him so persistently to return to the dark danger, again and again?

The next morning, sweeping the floor, I found Prionus dead beside the pine cabinet.

"It's dog days already. Snakes are blind. Their eyes are milky-looking," announced Dick at supper.

In the afternoon he and Joe had discovered a snake curled up near a rathole in the barn. Snakes are valuable for the quick and permanent way in which they take a rat's mind off his worries, so the men tried to shoo the snake down into the rathole. He uncurled his upper half and waved it back and forth, looking blindly out of his milky eyes. But when they set up a makeshift trough ending at the rathole, the waving head discovered it and the whole snake quickly slithered down into it.

A barn in late midsummer is as different from a winter barn as the two kinds of weather are different. Now in the barn only a few calves and ponies occupy the stalls, in company with the steer being fitted for the county fair next month.

In the airy, dry coolness they stand thoughtfully eating hay, now and then drinking water from a tub which is hose-filled from a pressure valve at the front of the barn. The water comes down to the valve from the big pond back of the barn. There is no large coming-in at night now or going-out in the morning, no pushing, shoving by authoritative, bigger cattle. A gentle breeze floats in through all the broken windows, barely stirring the loose straw on the lime-sprinkled floor. The barn smells quiet and clean and has the feeling of something that happened long, long ago.

IN THIS ERA when television threatens conversation with extinction, Ripley Renault is a comforting bulwark. His conversational methods are typical of farmers who leave the farm to take a job that keeps them in daily touch with farmers. Ripley drives a gasoline truck. He stops here every week to fill the gas tank.

It was Ripley who told me to hang a banana stalk in the henhouse to get rid of mites, and that a cow with "holler tail" must have her tail split and packed with salt. Once he killed a copperhead snake in the yard for me.

The thoroughness of Ripley's probing conversation is characteristic of the way farmers make sure you understand what they're telling you. He gathers up whatever news there is along his route and delivers it to the next place. He is instinctively a dramatist. Ask what's new and he says casually: "Oh, nuthin' much, I guess." After a pause there is an irrepressible flash in his eyes and he asks: "Reckon you heard about Bartholomew Duncan?"

"No."

"Hung hisself in the barn last night." Even though you never knew Mr. Duncan, this is news. You make a mistake if you admit you never knew him. Ripley will introduce him to you. "Oh, yes, you knew him. He comes to every sale. He's Bill Bensen's brother by his first wife."

If this seems complicated, you can ask (and later regret): "Where did he live?" Ripley will tell you if it takes all day.

"You know that big white oak where high water washed up all the cornstalks under Beanblossom bridge last spring?" You don't, so Ripley starts nearer your farm. "Well, you know where Aunt Libby Wright used to live, up there on the hill where the old washing machine set out, painted blue and filled with some kind of red flowers?"

This also is beyond the realm of your acquaintance, so Ripley, pitying you, moves closer. "Surely you know where Mable Howes had her strawberry patch when she was married to her first man that drank up the farm?"

Ashamed to admit further ignorance, you throw in a wild guess. "Where the sorghum mill was?"

"No, no. That's Fox Hollow. Bartholomew lives in just the opposite direction, over past Gobbler's Knob. You know the Crossroads school where Tom Peters always pastured his milk cow?"

Unless you want Ripley to start at your front porch and describe all the way from there to the Duncan place, you'd better force a gleam of comprehension into your eyes and say you do know. Be convincing, for Ripley is not easily fooled. From this point it will be possible for him to take you along the road to where, at last, you see poor old discouraged Mr. Duncan hanging from a rafter of his barn—no, by this time they've taken him down and there are friends and neighbors every place. You recognize some of them. But don't make the mistake of asking Ripley about the ones you don't know. He'll tell you.

NOW IN the last days of July autumn insects have already started singing. Crickets appear determinedly in the house, and jarflies last week began their shrill winding-up outcry.

I cannot bear to give up summer yet. I am glad sultry August is yet to come. It is a comfort to hear the bobwhite calling from the road's corner down near the red lilies. Almost a month after wheat cutting, the bobwhite's song is as summer-filled as it was a month before the wheat ripened. For this comfort, bobwhite, I shall remember to put extra amounts of cracked corn for you down under the maple trees this winter, when you will walk through snow to reach it, leaving your pronged snowshoe tracks in the deep white drift.

AUGUST

[AUGUST]

❦ ❦ ❦ ❦ *INDIANS* knew this month as the green-corn moon, an appropriate name. For now field corn is sure of itself, and late sweet corn is ready to be roasted in its tender inner shucks, or boiled six minutes in salted water, then salted and peppered, bathed in butter and bitten into.

In the green-corn moon there are days when you almost feel you could cook the corn in the steaming August day itself. By noon the days are parboiled, but the early mornings are dew-drenched and cool at the rim.

"If you can just manage to get a good night's sleep, you can get through these hot days," said Dick, lying down on the cool floor for a short nap before going back to the field after dinner.

In the green-corn moon ironweed blooms, elderberries darken. In the morning hours human ambition puts out new hopeful growth like the sprouts of cabbage on old stalks from which the heads have been taken and the new blooms and thin bright bean pods on old bushes in the frayed gardens.

Now that secrecy no longer matters, the summer dwellings of birds are revealed in tall weeds. There also are disclosed the "rabbit sets," flattened, grassy places where rabbits sat and pondered their problems—and, if once seen pondering there, never returned again.

In the green-corn moon, insects sing of autumn and the fuel-oil man sings of winter. But even without these, I know summer is crumbling away. When Dick told me he had been asked to judge the floats in the Greene County Fair parade, I knew summer was not long for this world.

For August is the month of fairs, county and state. By the time these all-absorbing events are over, corn is ripe enough to put into the silo, school is about to begin, and summer is truly a thing of the past.

The delight of fairs carries a farmer painlessly past that acute point where summer departs and autumn is there.

It was late Saturday evening. Outdoors the week had come as close to an end as it ever comes on a farm. In the kitchen a kettle of apple butter was cooking down to a glazed thickness, sending up tall hot spurts above the kettle and filling the house with a spicy good smell. Sunday's angel food was cooling upside down in a pan, and a ham baked to dry pink tenderness was almost cool in the red-enamelled iron roaster.

I poured a teakettle of hot water into a gallon crock of freshly clabbered milk, stirring it while the curd thickened, separating from the whey, and then put it to drain in a cone-shaped colander for cottage cheese. On the kitchen cabinet, waiting to be carried into the dining room and stored in the cherry corner cupboard there, were seventeen glasses of fresh grape-apple jelly.

In the kitchen the sense of comfort was made more complete by the rare fact that the house was clean all over.

"Something smells awful good in here," remarked Dick, coming in with the evening paper open in his hand. I looked quickly at the headlines; another Polaris had been successfully launched; the United States had spoken back sharply to Cuba; an astronaut was expressing his readiness to start to the moon in less than two years.

The apple butter spurted up again, splashing the newspaper. As I turned to pull the kettle back from the gas flame, I realized how much the glazed, spurting surface of the apple butter resembles the cratered far-off face of the moon as shown in the photographs displayed in the observatory at Mt. Palomar which houses the giant Hale telescope.

Remembering Palomar, which I had visited earlier in August, I said: "Palomar is where the United Nations ought to meet."

To eat a narrow lunch of cheese and crackers and

Thermos coffee at the small tables enclosed by giant mountain oak and pine, where birds flit freely, singing, and squirrels do the quarreling. Then to walk out of that enclosed coolness into the stoned August-hot parkway and then along the shrub-lined walkway to the Observatory itself, up its hard wide steps into a room made momentous with photographs taken with the giant 200-inch lens; to read the truths briefly summarized there; then to stand spellbound and think of one's self, and our nation, and the whole race of man in the boundless terms in which Palomar sees them, the whole earth with the United and ununited and disunited Nations held restlessly within it, the whole earth so small a dot in time and space—only a part of one small galaxy. To realize that there are countless galaxies; to see photographs of galaxies that lie trillions of light years beyond us. One reads again—to make sure he read it right—that one light year is the number of seconds in a year multiplied by 186,000, the number of miles per second light travels. One is first overwhelmed, completely silenced, then, finally, comforted. In so boundless a concept, what nation or civilization would possibly make a difference? Then follows quickly the greater truth, which gives an individual the courage to live or die for what he believes in—surely so boundless an event as creation must have had also boundless design and purpose and designer, in which everything, from the smallest grass blade to the greatest human soul, has been taken into account and assigned a part which is essential to the whole great Galaxy.

Fall mowing

ONE OF August's luxuries is the mowing of pastures. Not for hay, just for neatness.

"Doesn't make a dime," said Dick, who was so eager to get at the mowing this morning that he declined Warren's invitation to go along to the stockyards. "It's what it does for the spirit all year. Even under the snow—a late-mowed field has a comforting look all year."

Then, besides, what's mowed is done. It doesn't have to be worried about, left to cure, then raked up into a windrow and baled with the continual hope that it won't get wet. If the blue-black threatening sky breaks to pieces and falls, cooling the hot air and drenching the ground and all the mowed stuff on it, it doesn't make a bit of difference.

ALL SUNDAY afternoon the whole house smelled like a bakery. Carol was making Swedish tea-rings, trying to get a perfect one for her 4-H entry at the county fair the next day. Behind every 4-H exhibit lies the ghost of many hopeful repetitions. By the time the tenth tea-ring was thinly frosted and made exclamatory with the scarlet glisten of maraschino cherries, I was beginning to think that maybe the national 4-H motto, "to make the best better," ought to be, instead, "to make the best do."

The 4-H effort is not perfect, but in the modern pattern of farm living it is as important as the farm tractor or the kitchen stove. It requires unstinted generosity of time and faith from adult leaders; sometimes competitions bring deep heartache to a teen-ager. One danger is emphasis on winning more than on participating. Sometimes parents or supervisors push a promising child too much for his good or the good of others. The dangers, as well as the fine opportunities, of 4-H have been summed up ably by Purdue's Edna Troth, one of the foremost directors: "The blue-ribbon boy or girl is more important than the blue-ribbon project," she says.

There are valuable scholarships and honors to be won. The training in leadership and good-followership is as important as the skills achieved.

A strong 4-H club has almost the same spirit as a big family; its members are loyal, generous, candid; they quarrel; they work hard; they respect and help each other. For farm youth the 4-H club is the foundation of social life, with skating, swimming, caroling parties, hay rides, a talent show, a trip to a city amusement park, business meetings well conducted, judging contests; demonstrations of skill at club meetings and in district's competitions; a dress revue at which the sewing 4-H'ers model their projects shortly before the County Fair.

The 4-H club is a conveniently organized group for community chores, too, such as picking up beer cans and trash from the county road in the spring clean-up.

Parents are inextricably involved, as leaders, sponsors, or applauders. Jean and Russell Morgan are the leaders of the Maple Leaf club in this community. Their four daughters are immersed in 4-H'ness. At the Morgan's house, an ordinary Sunday dinner has the air of a 4-H demonstration.

"Sometimes," said Dick, "I think instead of 'head,

heart, hands, and health,' the 4-H pledge ought to mention 'hide, hair, horn, and hoof.' "

On the Monday morning of fair week, however, it ought to be simply: "Hurry, hurry, hurry, hurry."

By the time Jean arrived, with her station wagon already full of 4-H'ers and their fair entries, Carol had her tea-ring transparently wrapped. The polished cotton party dress, for which she had already won a blue ribbon in the dress revue, swung from its hanger as she hurried out. "I'll be satisfied with red on the tea-ring," she cried, "but I've just got to have blue on my dress!"

Right after the station wagon left, Warren's truck sped past, headed for the barn to pick up the four yearling Shorthorns Dick and Joe had got ready for the fair. The calves had been washed the week before and freshly Scotch-combed that morning. Their thick dark-red hair curled up in a fine fair-do. The show box had been packed. A pitchfork, a broom, water bucket and sponge, fan and radio, halters, a camp cot and bedding, three bales of the greenest hay and six bales of bright yellow straw (especially saved in June for this purpose) were ready to be loaded into the truck. Joe came to the house to get his billfold and tell me good-by. "I'm going with Warren. We'll stop and get Henry and his Jerseys and stuff. I'll see you at the fair!" He was as gay as sunlight dancing on water.

The fair would not be officially opened until that evening, but all exhibits had to be brought in that day. The food concessions were already operating. There would be sandwiches, hot coffee and cold drinks, and ice cream cones available. The carnival would be getting set up beyond the agricultural buildings. Trucks would be coming in all day with livestock and farm machinery and home equipment and the makings of a fair. The preliminary fair activities are as gay and busy and revealing as the days when the livestock is judged, the 4-H ribbons awarded, or the night shows come on. The greased-pig capture, the hog-calling, the afternoon balloon ascension, the night rodeo, the horse show, the donkey baseball game, the diving mule, midget-car races, the 4-H king and queen crowning, the parade of champions would follow on schedule, but no real fair addict could miss the preliminary. Dick couldn't. Besides, he was president of the fair board.

I began to wrap the nine unchosen tea-rings for the freezer—because there is a limit to the amount of tea-ring a family can stand, even when loyalty outruns appetite.

Dick came into the kitchen. He looked like a holiday about to dawn. "I'm going to take Joe's 4-H electric fan in the car," he said. "It has to be in by 10 o'clock. I'll probably stay." "Probably" was a blue-ribbon understatement. This is his idea of the farmer's perfect vacation: your own county fair all week, preferably with some good cattle exhibited; one or two days at fairs in all the counties you have lived in before; one or two in counties where you have good friends exhibiting; a few at fairs that offer outstanding horse pulls or tractor pulls; then two or three at your state fair; and around Thanksgiving, a day and a night at the International Livestock Show in Chicago. What more could a farmer want?

Just this: "One evening this week I want to see our county fair with you."

"I WAS just walking through the dairy tent," said Dick, "and I noticed the people crowding around toward the corner. They were standing three deep and watching something intently. One man was wearing shorts and a plaid sports shirt, and he seemed almost spellbound."

So of course Dick pushed forward curiously, too, and was tall enough to see over and between heads.

"It was only Calvin milking that big black and white Holstein of his. He was milking from the left side, but I knew that crowd wouldn't realize there was anything unusual about that. It was just that they were seeing something they didn't see often. Probably some of them had never seen a cow being milked, before."

NO MATTER how modern a county fair is, there is a vestigial quality about it that causes me always to wish I could show it to some ancient farmer—Mayan, Aztec, Roman—or some land-worker from some other vanished civilization.

I believe the most ancient farmer would quickly feel at home at a modern fair. He would probably enjoy the foot-long hot dog or the ubiquitous hamburger sandwich and soft drink, the wad of cotton candy, the caramel apple. The carnival midway, with its lights and music, fortune teller, wild rides, and tests of skill, would delight him.

Smugly, modern farmers would like to think the ancient one would be astonished by the high glossy quality of fruits and vegetables, or the deep, fine fiber in the flat-sheared fleece, the chest depth and pendulous udder of modern beef or dairy cattle, or even the gleaming plumbing in the commercial tent. He might, at first. But even without being able to speak a word of our language, he would soon feel at ease, for the basic spirit of the fair has cropped up in all civilizations—farmers celebrating their harvest, and townspeople, in recognition of the importance of agriculture to their civilization, celebrating the farm harvest with the farmers. If the townsman can at the same time interest the farmer in a little of the town's exchange, so much the better. The commercial tent, especially to a farmer who is interested in a new freezer, pump, tractor, or electric organ, or whatever, is as absorbing as the agricultural exhibits.

As we approached, the fair showed up first as a pool of light in the darkness; the upper arc of the Ferris wheel was a jeweled bracelet on an arm flung up against a dark sky. Nearer, we heard the fair sounds, music from the midway competing with music from the fence-enclosed arena. When we slowed at the gate to let the guard see the pass fastened to the windshield, we could smell the distinctive smell of county fair, combining odors of popcorn, cooking food, burning tobacco, sweating people with fan-cooled, well-groomed livestock, cleaned barn floors, and a field trodden to dust by many feet.

Dick turned to me, smiling: "The pulse quickens, as your father used to say. What do you want to see first?"

Already I knew Carol had won red on the tea-ring, blue and also purple (meaning it can go to the state fair) on the dress; Joe's fan had a blue ribbon; and the cattle had some blue and some purple ribbons. Our rallying point throughout the evening would be the beef barn, but I wanted to see the 4-H and the women's exhibits so I could compliment everyone on his honors. I wanted to see the flowers, the handmade rugs, the lovely serene quilts hanging from their racks, and the incredible things people crochet. Much of the handicraft is pure lavishment of imagination, with no hampering of function or beauty. Much of it, surely, is put together merely to show that anything can be used in some way, and much of it deserves a ribbon for proving this.

I always want to see the poultry. It makes no difference that guineas, chickens, ducks, geese, turkeys, and rabbits look the same and smell the same every year; I look forward to every one of them. I want to see the goats because their limpid, patient, intelligent eyes remind me happily of "Bo" Bowman's goats, Mike and Ike, which we borrowed one summer with the small farm wagon and double harness made for goats. I want to see the dairy cattle and the beef cattle because we have Maple Grove neighbors exhibiting both. I want to see the wonderful hogs, black, red, white, spotted, belted. They are huge and clean; their feet are polished; they look downright noble, a genuine achievement for a hog, although hogs are, admittedly, the most intelligent, certainly the most shrewd, of all farm animals, including horses.

In our county fair there is no horse barn. Horses are brought in only for the night shows and 4-H exhibits. Some county fairs have an overproportion of horse interest. The atmosphere around a horse barn is not at all like the democracy of the cattle or hog barn. There is a distinct air of snobbery at a fair's horse barn; whether it originates from the horses or the owners is not important.

In the barns, already, ribbons were swaying from above the stalls, stirred by air created to make the places endurable for the farm boys who sleep there at nights during the fair and care for their stock.

In the sheep barn that evening there was a weaving exhibit, with loom, spinning wheel, and hand-woven fabrics. In the afternoon there had been a sheep-shearing contest there. Now in the barn some of the sheep were blanketed to keep their flat-tops clean.

"Sheep have the most human-looking faces, don't they?" I said, and Dick nodded. "I used to have one named Ethel because she looked like the hired man's wife, and one named Lorraine." A faint sound of laughter in the arena was blown into the sheep tent.

"There was one old ewe," said Dick, pronouncing it "yoe" as farmers do, "that we had when we lived in the big brick. She swallowed the chain on the end of a singletree. When she got to the wood, it was stiff, of course, and she began to back away from it and pull the chain out. But the hook caught in her cheek and came out through the skin. She was the silliest looking thing when I found her, just standing there with that hook sticking out of her face. I unhooked her, but it left a hole and always after when she chewed grass the juice ran out of her face."

We started to the horticultural tent, because I particularly enjoy the 4-H forestry exhibits of leaves, woods, and seeds and the 4-H entymology exhibits of moth, beetle, praying mantis, dragonfly, and other familiar and delightful nature friends. My only regret is that these friends are dead. We made slow progress across the fairground; in fact we did not really get all over it, because we kept meeting friends and stopping to talk. This is one of the happy reasons why a county fair has to last a week. You can't possibly see everybody in less time.

After a day or so, everybody is tired but nobody wants to miss anything. On the last night, when finally the trucks can come in and take the exhibits away, everyone is so tired all he can think of is how glad he is to go home. But all the next week, when friends and neighbors meet, the first question will be: "Well, what did you think of the fair?"

BETWEEN 6:00 and 7:00 in the evening the cows in the hill pasture and the calves in the barnlot remembered their mutual dependence and began to bawl. The calves huddled in a corner on their side of the fence; the cows started from the pasture. I was thus reminded that the rest of the family had gone early to a fair, and it was my chore to let the cows in with calves.

It had been a hot day, but now, just before sunset, the air was cooling. The fragrance of Jimson began to drift out from the tall weeds near the fence as I opened the gate.

A big roan cow was already there, bawling to the calf, which bawled back from his side of the fence. When I opened the gate, they ran eagerly together and the calf began to suck out the warm foamy supper the cow had spent all day preparing. Foam covered his lips and chin, overflowed the long trough of his tongue, and dropped like big white marshmallows upon the dry ground.

The smell of warm milk and the sight of this deliciousness goaded the other calves to louder effort. The cows came downhill singly, or by two or threes. Some came slowly; some ran. All came bawling.

To hear a cow bawl is only a small part of the drama. To watch her is much more. She lowers her head and stretches her neck forward so that her lower jaw is on a line parallel with her back. Then she swivels the rest of her head straight upward, thus creating a deep channel emptying from her soul to her open mouth. It gives her voice splendid carrying power; even a small cow can make herself heard at a great distance. If a farmer goes to bed having forgot to turn her in with the calf, she can penetrate his soundest sleep with a few effortless blasts.

There is always one cow who wants special attention, and she will get it by being inconsistent. Having reached the very edge of reunion with her calf after bawling all the way down to the fence, she suddenly decides she is in no hurry for him. She turns and walks along the fence, purely so that the farmer will pay her the special attention of going after her.

Why this nightly matching of wits, and watching the simple daily reunion of cow and calf, has such appeal, no farmer can say, but almost any farmer will say that one of the pleasantest times of his day is the time when he turns the cows and calves together and then just stands leaning on the fence, watching them.

WHAT A BLESSING it is that even in this age of efficiency there are still some small, inexacting household chores that have to be done by hand.

In moments of sudden deep grief, or the overburden of anxiety, or unbearable joy, such simple things as pressing a pan of cooked apples through a sieve or sweeping a kitchen floor, or making up a bed with fresh, clean sheets, can be genuinely steadying.

I'm vigorously in favor of having all the modern mechanical conveniences you can get, but somewhere I want to keep a few unimportant hand chores, so that I can get them out in times of need, like the candles kept in the buffet drawer for the infrequent but inevitable times when the electric current fails.

FOR THIS one night of dramatic bloom, Ola Robinson had tended the tall, ungainly cereus for seven years.

She had carried the two-gallon flowerpot into the house

for winter, fed the plant a liquid plant food, watered it when she thought it needed it. In summer the cereus stayed in the back yard, among other pots and kettles of growing plants. The Friday before it bloomed it had gone through a windstorm that made a shambles out of Ola's vegetable garden.

Two weeks earlier, when the small, red, oval buds emerged from the narrow edge of the long, rubbery leaves, Ola had alerted her farm neighbors. When the buds turned reddish-brown and pushed upward into the shape of candle flame, she knew they were going to open that night.

She carried the plant to the open front porch of her big, weather-beaten colonial house and turned a strong light on it. She brought chairs and rockers out into the yard and built a gnat smudge. She brought out a reading glass. There were twenty-five people there when we arrived; more came and left throughout the evening. It was open house for the cereus.

At dusk three flowers were opening slowly, white as new snow, four inches across and shaped like a water lily turned on its side. By midnight, fully expanded, they would be seven inches across. By morning they would be closed, pointing straight down on the bent-over stems, and done blooming.

Now each flower was a tiny immaculately white stage, and inside of it, the scenes were changing. Looking far back into its tubelike depth, you could see what people call the cradle and canopy, and the star. Actually, these are made by the stamens, like heavy white threads, that rise, droop, divide, and reach forward, ending in powdery knobs toward the front of the flower. At the front of the flower the pronged tube makes a distinct star.

As the blossom slowly expanded, the cradle scene changed; now it became a group of white-robed people standing at a solemn feast. The central figure rose a head taller than the rest, and anyone observing was reminded of da Vinci's *Last Supper*. This scene changed almost imperceptibly, the way nebulous clouds change, and finally became two white-robed choirs.

The fragrance accompanying this impressive, Biblical-minded flower is light but subtle. You have to bend close to catch it. One guest thought it smelled like May-apple bloom; one compared it to the fragrance of Jimson. All the night-blooming cereus seems to lack, in its drama, is the ability to make solemn, quiet music.

IT WAS Sunday morning, and the kind of day that makes poets out of hill farmers. Dick laid down the piece of toast on which he was about to put a forkful of red raspberry jelly and looked through the window at the hilltop cornfield. "A field of corn is a beautiful thing at any time of the year," he said softly. "My, I'm glad I got the ragweeds moved from in front of it yesterday!"

The ripening field was like a big layer cake: first a layer of thin brown, then one of rich green, then a final layer of tassels the color of toasted coconut. Above the cornfield hung the clear untroubled Sunday morning sky. It was pure poetry to look at.

Corn is the measure of the farmer's skill and soil-building practices. If it does well, the farmer knows he and the field have done well, whether the price is 35 cents a bushel or $2.27.

Yesterday, when I was picking the last of the shellout beans from vines twined around the first two rows of corn, I noted that the cornstalks are higher than the transom above the kitchen door. On each tall stalk is one large ear; and lower down, a small one. To produce so much stalk in order to get just one ear seems extravagant. If a farmer puts the corn in a silo, he gets succulence from both stalk and immature corn ear. Corn is ready to put into the silo when the top ear is just past the roasting-ear stage.

In the corridors between rows of corn, late-sprouted cocklebur and Jimson are thriving, and wild sweet potato lifts its milk-white cups there. All these came up after the corn was laid by and will still have time to ripen seeds for troubling the field in another year.

The cornstalks are not troubled by these squatters at their feet, sharing their food and drink. Their concern is how to remain upright, and they have prepared for this in a thrifty, farsighted way. From the first joint above the ground they have sent out heavy short stubs that look as if the cultivator shovels had come too close and cut off the roots. These are incipient brace roots. If the stalk is weakened by weed-spraying or threatened by strong winds, the braces will grow and strike down into the ground for firmer anchorage. This makes corn seem almost to possess an intelligence.

In daytime now the tassels shake out their sweet perfume and scatter it into the air. At night the heavy, sweet, unripe perfume of corn gathers like fog in low places; if you drive past a cornfield in the dark, you know it is there because you can smell it. And in the sultry nights of August the stalks snap and rustle as they grow.

AT McCormick's Creek park in Owen County yesterday we watched a line of eighteen adventurous riders come out on their rented horses to ride over the park's long narrow bridle path through the deep fine woods. There old John McCormick used to ride his work horses in from a day's plowing, past the little farmhouse on a knob beyond the pear orchard, past the springhouse, and on out to the big barn, which had an eighty-foot threshing floor with no supporting posts to interrupt the horses as they ran around and around, threshing out the grain.

We had already met some of the park concession's horse-drawn rigs. There was a big, slow-walking roan pulling an open-top surrey, in which rode a man and three happy little children. A trio of young women were starting out in a two-seated open-top surrey pulled by an excessively trustworthy mare. Two girls were sitting in the front seat holding the lines lifted up and taut; the third girl leaned forward anxiously from the back seat and asked: "Do you think you can guide it?" A little distance behind them came a buggy with some small children, a man, and a woman in it. "Be sure you pull the right line," the woman advised the man as the horse clopped inchingly past.

Farm-safety week is over. If it had lasted any longer, everything on the farm would have been done for.

Farming is listed as the third most hazardous occupation in the United States.

Some accidents are due to carelessness; the penalty for them is swiftly imposed and relentlessly collected. A corn-picker, for example, imposes a savage penalty for carelessness and never suspends sentence for good behavior afterward.

The freak accident happens under unforeseeable circumstances. It is for freak accidents farmers carry insurance. One like this overtook Dick early in farm-safety week. Wearing a hat, he looked up into an apple tree, as he walked under. A green apple, having started to fall at the right second, met his face at exactly the second necessary to black his eye. This wasn't so bad, although a black eye appearing suddenly on a farmer's face can cause his wife considerable embarrassment—especially if the neighbors just look and tactfully avoid mentioning it. All the farmer needs to say, casually, is: "This hot August weather is awful hard on a woman's temper."

Thursday afternoon when I went to help Grace can beans, Dick was getting the mower ready to mow the pony lot. "A mower takes the hardest beating of any tool on the farm," he said. I believe it. For every hour mowing in the field, it demands three hours of repair, filing, sharpening, adjusting.

"Be careful with that pressure canner," Dick said, "and drive carefully, too. Remember this is farm-safety week."

When I returned at dusk, the pony lot was neat as a new flat-top haircut. I started supper.

When Dick came in to supper, his cheek was criss-crossed with band-aids under which blood had run down and dried. His jaw was swollen. His eye was black again. In adjusting the cutter bar, he had loosened a double nut, releasing a heavy spring and thrusting out a heavy iron bar that knocked him several feet beyond the mower. An inch and a half lower, or any nearer the eye, could have meant a disastrous accident. "And not a car came along for two hours," he added thoughtfully.

There's a farm superstition that mishaps come in threes. "Something must have it in for this family," I said. "I bear you no ill will, but maybe it's my wifely duty to strike a third blow, just to make sure it's light."

But these things seem beyond human control. The next morning the telephone rang and a distressed voice said: "Your sister Grace has just fallen at folk-dancing class and broken her arm. Can you come and take her to the hospital to have it set?"

A POKEWEED PLANT came up this spring near the red raspberries, and I left it. Now it is eight feet high and its first berries are full of juice, inviting thoughts of making ink or dyeing cloth. The color of ripe-pokeberry juice, dark, filtered, purple-pink, is absolute—like a measure, like the unvariance of twelve inches to the foot. The berries are poisonous for people to eat but safe for birds.

Pokeberry has other names, pigeonberry, garget, scoke, coakum, and inkberry. It is a perennial. People can eat its early spring shoots for greens, or like asparagus, but its full-grown leaves are poisonous. Its small, inconspicuous flowers, set wide apart in a long cluster, seem like tailor's tacks put there temporarily to indicate where something else is to go. Little, flat, green berries, pumpkin-shaped, quickly replace the first flowers. Like the orange tree, a pokeberry bears blossoms and green and ripe berries all at the same time. The stems and tall stalks are pink, as if the berry's stain had soaked down into them.

The stalks die in the first hard freeze, but the roots survive deep in the cold earth. The berries have already been eaten by birds, long before that. Winter rains and snows wash off the soft dead leaves and darken the bark. By the next spring the pokeweed's old stalks are white as old bones, the bark peeling off like strips of rag. Inside, the stalks are filled with transparent disks. You can read through these disks if for some reason you want to. The old stalks break off readily then; the new sprouts are there, needing the space.

Nature has the most admirable way of handling such things. To her, death is not the opposite of life but an essential part of it, and nothing is ever wasted or lost.

POOR ALBERT is dead, and the farm is therefore the poorer. For Albert was the clown, and a farm needs a touch of humor as truly as bread needs a touch of salt.

Death of a clown

Albert was a goose and probably destined, from the beginning, to be a clown. The preposterousness that characterized him and endeared him to all of us began even before he was hatched from the egg.

A banty hen wanted to set, so Joe put four goose eggs under her. A banty, being much smaller than a normal hen, is very much smaller than a goose. In comparison to goose eggs, a banty's eggs seem hardly larger than gooseberries.

Sitting, elevated and slightly bewildered, on the four goose eggs, the banty looked preposterous, but she sat willingly. Albert was in one of the eggs, although neither the banty nor we nor even Albert himself realized it at the time.

The banty was faithful and after four weeks three fluffy, appealing, yellow goslings hatched. Two met with fatal disaster while still in the downy stage; the fourth died in the shell. This left only Albert. Alone and lonely, he was brought up by the conscientious little mother. By the time his bright yellow down had paled to white and was beginning to be replaced by feathers, he was much larger than the hen but had not yet attained the proud aloofness of an adult goose. He was gangling and awkward, fearful of being left alone. When he ran after his mother, he frequently ran into her and knocked her over. Walking or running, he went about with his neck stretched far out, head down to shoulder level, eyes bright. And he whistled.

He tried to take up with the other poultry, the many families of Bantam chicks, the various sizes of full-skirted young guineas, the mallard ducklings (three sizes from three different hatchings). Baffled by his whistling, they were all afraid of him. In his effort to be friendly, he whistled the more cordially.

No matter what he was doing, Albert looked ridiculous. It was his one talent; the gift of laughter was his one gift to the farm.

When he ran through the green weeds, like a white shuttle through a carpet loom, there was laughter in watching him. We all liked Albert and petted him.

When he ran with the chickens, he was a land bird. From running after the mallards he discovered the delights of water, and spent some happy time in the shallow water remaining in the terraces after rain. He emerged from his play there with his pumpkin-yellow legs and webbed feet washed clean.

The last time I saw Albert he was standing still, holding up one yellow webbed foot, and when he put it down to take a step, he limped. Two mornings later, only one pumpkin-yellow leg was there, in Albert's coop. A sad and angry search located poor Albert's partly eaten body buried lightly under loose earth at the big pond. But let there be no tears. Laughter was Albert's mission; let a smile be his epitaph.

OPPORTUNIA is fat now. In the animal world fatness is evidence of easy living, though the question whether easy living is more desirable than freedom with hunger is always debatable.

A skunk makes a charming pet

Opportunia is a young skunk, picked up by Joe and Henry a few weeks ago when she was alone, thin, a weaned baby. The boys put her into a cardboard box, against which she used her immature powers of defense. Scent-throwing is a skunk's only weapon, and less powerful in a baby than in an adult; still, nobody would have had any trouble guessing what was in the box.

Henry was going into Navy duty within two weeks, so the baby skunk became the charge of Joe and nothing could have pleased him more, except more skunks or animal pets of any kind. That first evening he stroked her white-striped, black body gently with a piece of wood. The next day she accepted the same stroking from his hand. After a week he could safely pick her up and cuddle her against his shoulder. Now she likes having her narrow, triangular head scratched behind the ears, a delight second only to food for wild animals, when once they become able to bear the touch of the human hand.

Opportunia likes cereal, milk, green beans, cooked meat; she will eat anything she is offered, except potatoes or cantaloupe.

From the white cap at the back of her head, two long white stripes go down her back. Her black tail is tipped in white as if she had walked too close to a freshly painted white wall. On her nose is a match-sized white stripe. Her eyes are dark, shiny, inscrutable, her nose like puckered black rubber. She looks just like Petunia, the adult descented pet skunk given us by the Forest Payne family four years ago. A de-scented skunk smells always somewhat like bread that has been dried very hard and pale-brown from being a long time in a very slow oven.

A skunk, being wide at the back and narrow at the head, cannot be harnessed. Petunia had only to step backward to be out of it. Because skunks are nocturnal creatures, she slept in daytime and at night, kept in the house, was restless. We heard her hard little black toenails striking the linoleum as she ran about the house. When we let her out, she disappeared but at night returned to the corncrib. When the children discovered her tracks in the dust there, they baited a wooden crate with food and recaptured her. After that we understood each other. We gave her her freedom. She came at suppertime and pushed open the back porch door, as Rose did, and scratched at the kitchen door to be let in, petted, and given supper.

Tragically, we did not know that skunks are particularly vulnerable to rabies, so we failed to give her a shot when Rose had one. Her death was an acutely painful experience for all of us.

Rose was never enthusiastic about Petunia. At first, recognizing her as one of the creatures whose weapon is never to be challenged, she avoided her completely. When she learned that Petunia was associatable, she decided all skunks had reformed. One of Petunia's free-lancing guests enlightened her on this point. Nauseated, Rose rolled in the grass; she foamed at the mouth; rolled in horse manure, her favorite perfume; she went to the garden and rolled in loose, tilled earth. It was her bath salts, her towel, her everlasting refuge. After that, as far as skunks were concerned, Rose washed her paws of the whole breed.

Ever since we lost Petunia, we have wished for another pet skunk. We would like to keep Opportunia and have her de-scented. But Opportunia belongs to the wild world. Freedom is her right, and before school begins next month, she will be given opportunity to make a choice. It may be that she will choose freedom, above security.

A BLUE SAUCER fell off the table and broke this morning, and although my housekeeping philosophy has long since got beyond the place where such things are disasters, I regretted it. It was a saucer from our best set; it is pretty and has been used at many of our happiest gatherings. A cup fell at the same time, but I snatched determinedly as it fell past and thus saved it.

It seems fitting to bury the saucer respectfully in the yard instead of throwing it away with the debris of daily living, such as tin cans, catsup bottles, broken glass jars, worn-out pans.

I think burying is not an unusual disposition to make of broken dishes on a farm. On all the farms Dick and I have lived on (this is the sixth), I have dug up fragments that surely could not have got so deeply covered without help. Even where the rain washes most deeply, erosion could hardly have covered so well some of the artifacts I have found in digging down to plant shrubs.

Near the back step I turned up a small white ironstone

cup from a child's doll teaset. It was whole except for a chip on the rim. Once I dug up pieces of an extremely ornate vase, with china birds and flowers affixed to the outside. Someone must have felt either sharply unhappy or greatly relieved when that vase broke. On various farms I have unburied pieces of blue and white tureen, bits of pink plate, parts of old-fashioned china dolls with hair in shining, ceramic waves. I have found headless bodies of small dolls, and broken pieces of blue glass and green glass that I could almost have wept over. Once I dug up a small cut-glass salt dish and another time a small white saucer, cracked, whose mission was to hold the coffee cup while the cooled coffee was being swallowed from a larger saucer.

On the Johnson County farm, from which we moved to our present farm, I found a rusted button. It was embossed with the profile of Zachary Taylor and the words "Old Rough & Ready," so I assume it must have been worn by a soldier in the Mexican War when Taylor commanded the U.S. Army on the Texas border in 1845. He had earned that nickname from his activities in the Black Hawk and Seminole Indian wars several years earlier.

On this farm for a long time after a heavy rain we found old coins in the mud near the barn that later burned.

Little by little, in these vestiges of small household misfortunes long buried and forgotten, an old farm gives up the story of its former owners.

Some day perhaps another farm woman, digging down to plant a lily bed, will turn up the halves of my Enoch Wood blue saucer and stop to wipe the dirt from it. At noon she will show it to her family, perhaps even comment that "it was a pretty saucer; I'll bet that family had a good life here."

AFTER ALL the fairs are over, the fall festival comes on.

Held annually in small Ellettsville, the festival began more than twenty years ago as a homemade way to do honor to a particularly successful 4-H group. They had only home talent for entertainment, but it was fun preparing and doing it. They had some exhibits and a good time. Visitors enjoyed it, and the idea took hold like a hungry calf. People worked for it eagerly, with no thought of pay.

It has become a combination of old-time Chautauqua, old settler's week, and county fair. Paid professional entertainers perform on an open-air stage at the end of the main street (in front of an antique shop), but there is no charge to the audience. Bleacher seats and benches in the main streets are provided with no charge. The weekly Ellettsville Journal office, (festival headquarters) is on this street. So also is the town's one saloon, whose owner boards up his windows and goes on a vacation during the festival.

In the three days of festival, the town's customary population, around 900, swells to 12,000. Only neighborliness and good will explain how so many cars find room to park.

Each year's festival has a different theme—such as "Ain't God good to Indiana!" or "Back Home Again in Indiana," but the slogan is always the same, "Hi, neighbor!" You hear it continually, and see it pasted on the store windows.

There is a festival queen (the contest helps pay for the paid entertainers), and traditionally the Governor comes down from Indianapolis to crown and kiss the queen. At that time the festival president, editor Maurice Endwright, is in such a state of joy over the success of the festival that he also kisses the queen and would as readily kiss the Governor. The editor is a small, unpretentious, diplomatic, and greatly witty young man with gentle brown eyes, a shy, soft voice, and astonishing organizational power and firmness of purpose.

There is a parade on Friday afternoon featuring education and schools, but the highlight of the festival is the Saturday afternoon parade. On these two days you are likely to see any strange object, representing the past, present, or future, traveling along the roads toward Ellettsville, to be in the parade.

There is a reviewing stand beside the highway for the judges and the many celebreties who come: senators, editors, governors, favorite sons, Congressmen, Army and Navy brass. There are traffic police. Children run freely from their parents, being as safe as in a giant playpen. There are refreshment stands, exhibits such as fairs offer, a commercial tent—all set up in various streets in the little town. The grocery stores, dry-goods and department store, drugstores, hardware store, and the new post office try to do business as usual, but it's about like trying to can peaches while playing basketball.

It is a great event, but so much hard work that every year editor Endwright pretends gloomily he doesn't know whether the town can put on another. Why do we have a festival anyway, the workers ask; and nobody can really say.

ELDERBERRIES are ripe now along the country roads. The bushes bend forward, offering their clusters of black-purple fruit. They have no thorns, and the berries are easy to reach and can make palatable pies or jelly, but they are tedious to pull from their individual, rosy-purple stems.

They are inviting to the many birds that will feast thankfully there from now until after frost.

The blossoms can be used to make a pale, pleasant wine, somewhat like dandelion wine, but few farmers make it.

The berries have a distinctive but somewhat cloying flavor unless lemon juice is added. In pies one also adds butter, and possibly a sprinkling of cinnamon. Every one of these small berries (botanically a drupe, which means one seed surrounded by juice and is not as scathing as it sounds) will make a surprising amount of stain. On your fingers, on the white tablecloth, on the dishtowel, when the juice touches water, it turns an inky persistent blue.

But an elderberry hedge in bloom is a charming thing. It provides shelter and food for birds and fragrances for human pleasure. Wind goes through the delicate clustered blossoms and shakes out the fruity fragrance into the early summer air. When you first smell elderberry blossom, you know summer is there, and at that point it seems long, unhurried, unspent.

When the elderberries ripen, you see the outer boundary of summer, and you know summer is nearly spent.

THE BIG oak tree fell in last night's violent windstorm and thereby probably got more attention than ever before in its long lifetime.

People have driven along the road past it, have parked near it or under its wide shelter to go into the church, but everyone took it for granted, never really seeing it. Nobody realized how big it actually was until they saw it lying on the ground. When it toppled, it covered the yard with astonishment, on account of the amount of wood in it and the long logs in its mighty, outspread arms and the depth of hollowness in its top trunk.

It fell like a giant lying down to take a nap, and rested on its elbows, so that it did not crush the long iron gate in the old stone wall enclosing the cemetery, nor knock over much of the wall itself. The farthest limbs of the oak reached out nearly a third of the distance across the cemetery. Some of its small twigs and graceful scalloped green leaves broke free and flew even farther.

It will take the church's farmers several days with power saws and tractors to work up the old tree and take it away.

Nobody knows exactly how old it is. Nobody in the community is old enough to remember when the top blew out of it. The live bark and solid limbs surrounding the broken top gave it symmetry and a look of wholeness.

The top-breaking occurred so long ago that within the

hollow trunk the tree's own wood had fallen, rotted and crumbled. Rain and wind, assaulting the hollow, had pushed the crumbled dark wood down between the layers of old and new wood, so that when a veneer-thin strip was pulled from the jagged stump, the dark moist oak crumbs poured out like wet corn meal.

I went up and filled a gallon can with the crumbs, to bring down to the house and plant something in. Packed in with the crumbling wood were old pignut shells, like black rubber, and acorns left over from some squirrel's hoard.

It is a basic and comforting principle of nature that nothing is ever really destroyed or wasted. Damage to shape or form does not destroy the elements that were contained in it. From nature's viewpoint, the storm damage was not disaster; it was merely the beginning of another and different use of that material, another chore assigned to the components of the oak tree.

❦ THE SUN comes now at a slanting angle; early mornings are marked by an almost springlike excitement, as if the year might go backward and return to September by way of spring.

It is the end of August, the end of summer, the end of a book. But it is not really the end of a year, because a year is not really a circle. It is a segment on a spiral that repeats the same pattern year after year, but never quite identically, and never comes together in a final and hidden meeting place as a circle would.

August

The year is life itself, forever changing, forever leading on to something else. In each year we look for cherished and familiar yearmarks, and in finding these, we discover the necessity of the suffering the year exacts, the discipline it imposes, as well as the generosities it pours out to us.

We do not know where the curving, spiraled pattern leads, nor in fact whether it leads up or down. We know only that someplace on it there is an assignment for even the least among us, and that the fulfillment of that assignment is important to the pattern and ennobling to the individual.

A Note about the Author

RACHEL PEDEN (1901–1975) graduated Phi Beta Kappa from Indiana University in 1923. She wrote a column: "The Hoosier Farm Wife Says. . ." for the *Indianapolis Star* and "The Almanac of Poor Richard's Wife" for the *Muncie Star.* She also authored three books: *Rural Free: A Farmwife's Almanac of Country Living; The Land, The People,* about the family farm; and *Speak to the Earth,* about the virtues of farm life. Peden was chosen as one of the ten charter members of Monroe County's Hall of Fame.

A Note on the Type

THE TEXT of this book was originally set in MONTICELLO, a Linotype revival of the original Binny & Ronaldson Roman No. 1, cut by Archibald Binny and cast in 1796 by that Philadelphia type foundry. The face was named Monticello in honor of its use in the monumental fifty-volume *Papers of Thomas Jefferson,* published by Princeton University Press. Monticello is a transitional type design, embodying certain features of Bulmer and Baskerville, but it is a distinguished face in its own right.

For the 2009 edition, Adobe Caslon Pro, and ITC New Baskerville were also employed.

Original design by
VINCENT TORRE

CPSIA information can be obtained at www.ICGtesting.com
Printed in the USA
LVOW061837160113

315942LV00006B/517/P